FOUNDATIONS OF RUSSIAN MILITARY FLIGHT, 1885–1925

FOUNDATIONS OF RUSSIAN MILITARY FLIGHT, 1885–1925

JAMES K. LIBBEY

NAVAL INSTITUTE PRESS

ANNAPOLIS, MARYLAND

This book has been brought to publication with the generous assistance of Marguerite and Gerry Lenfest.

Naval Institute Press
291 Wood Road
Annapolis, MD 21402

© 2019 by James K. Libbey
All rights reserved. No part of this book may be reproduced or utilized in any form or by any means, electronic or mechanical, including photocopying and recording, or by any information storage and retrieval system, without permission in writing from the publisher.

Library of Congress Cataloging-in-Publication Data
Names: Libbey, James K., author.
Title: Foundations of Russian military flight, 1885–1925 / James K. Libbey.
Description: Annapolis, MD : Naval Institute Press, [2019] | Includes bibliographical references and index.
Identifiers: LCCN 2018051296 (print) | LCCN 2019000004 (ebook) | ISBN 9781682474327 (ePDF) | ISBN 9781682474235 (hardcover : alk. paper)
Subjects: LCSH: Aeronautics, Military—Soviet Union—History. | Air power—Soviet Union—History. | Soviet Union. Raboche-Krest'ianskaia Krasnaia Armiia. Voenno-Vozdushnye Sily—History.
Classification: LCC UG635.S65 (ebook) | LCC UG635.S65 L53 2019 (print) | DDC 358.4/03094709041—dc23
LC record available at https://lccn.loc.gov/2018051296

♾ Print editions meet the requirements of ANSI/NISO z39.48-1992 (Permanence of Paper).
Printed in the United States of America.

27 26 25 24 23 22 21 20 19   9 8 7 6 5 4 3 2 1
First printing

All maps courtesy of Robert W. Carberry.

TO JOYCE'S FRIEND
DMITRY AVDEEV

CONTENTS

LIST OF MAPS ix
ACKNOWLEDGMENTS xi

1. Preparing the Way 1
2. On the Eve and Start of the Great War 20
3. Fall 1914 Campaign 41
4. New Roles for Aircraft in 1915 58
5. Flight during the Great Retreat 73
6. The Height of the Air War 89
7. The 1917 Revolution Impacts Squadrons 106
8. Reds versus Whites 123
9. Aviation and the Civil War 143
10. Soviet Victories in 1920 and 1921 161
11. Aircraft Development, 1918–1924 177

Conclusion 192

NOTES 199
BIBLIOGRAPHY 221
INDEX 227

MAPS

1. Eastern Borders of Germany and Austria with Russia — 30
2. Black Sea during the Great War — 31
3. Russia's Great Retreat from April to September 1915 — 32
4. Baltic Sea Region during the Great War — 33
5. North Russia during the Civil War — 134
6. From Omsk, Siberia, to Ufa during the Civil War — 135
7. Crimea during the Civil War — 136

ACKNOWLEDGMENTS

I AM INDEBTED to numerous friends and professional colleagues who responded to my request for assistance in the preparation of this book. On one of my trips to Russia, Mikhail Baskov facilitated my research efforts and even drove me to Monino to spend time at the Russian Air Force Museum. Special thanks must also go to librarians at the Jack Hunt Library of Embry-Riddle Aeronautical University (ERAU) in Daytona Beach, Florida. The library has one of the more complete collections of aviation and aerospace materials. When rare or Russian-language books were needed, Sue Burkhart and Elizabeth Sterthaus efficiently secured them through interlibrary loans. Suzanne Eichler graciously answered my general questions about the library.

Many individuals took the time to read a chapter, and their reviews resulted in suggestions or corrections that improved the manuscript. They include Carl J. Bobrow, museum specialist, Collections Processing Unit, National Air and Space Museum; Stephen G. Craft, global conflict studies professor, ERAU; Tom D. Crouch, senior curator, National Air and Space Museum; James M. Cunningham, former associate vice president for academic affairs, ERAU; Glenn J. Dorn, global conflict studies chair, ERAU; Charles Eastlake, former aerospace engineering technology professor, ERAU; George C. Herring, alumni professor of history emeritus, University of Kentucky; Thomas B. Hilburn, former distinguished engineering professor, ERAU; Russell Lee, Aeronautics Department chair, National Air and Space Museum; Bruce Menning, Russian military specialist, U.S. Army Command and Staff College; Robert Oxley, former associate vice president for academic affairs, ERAU; and Col. Ted R. Powers Jr., former air science professor, ERAU.

Finally, I am grateful for the help in obtaining photographic images and maps. Kate Igoe, permissions archivist at the Smithsonian National Air and Space Museum, spent time identifying appropriate Russian photos available from the museum; Cindy Taylor, retired from the Boeing Company, resurrected useful prints that are one hundred years old; and Robert Carberry applied his skills as a cartographer to produce excellent maps that illustrate the book's narrative.

This book is dedicated to Dmitry Avdeev. As one of my students from Russia, he holds dual citizenship in the land of his birth and in the United States. Today Dmitry is a commercial pilot and an aviation expert, specializing in oversized and humanitarian air cargo transportation.

Note: some Russian dates are presented in the Old Style (O.S.) of the Julian calendar. In the twentieth century the Old Style was thirteen days behind the New Style (N.S.) of the Gregorian calendar. The United States and many other countries use Gregorian dates. Soviet Russia adopted the New Style calendar early in 1918. For the years 1917 on, the text uses the New Style calendar exclusively. Except for Russians who later became Americans, Russian names are transliterated from the Cyrillic to the Latin.

CHAPTER 1

PREPARING THE WAY

★ ★ ★ ★ ★

SERBIAN NATIONALIST Gavrilo Princep assassinated Franz Ferdinand, heir to the throne of the Austro-Hungarian Empire, when the archduke visited Sarajevo on June 28, 1914. As capital of a province illegally absorbed by Austria and ardently desired by Serbia, Sarajevo came to symbolize the nationalism and rivalry that helped ignite the Great War, later known as World War I. The Austrian declaration of war on Serbia, on July 28, set into motion the alliance system that pitted the Central Powers of Austria, Germany, and later Turkey against the Allied Powers of Great Britain, France, Russia, and eventually Italy. Over time ten other nations joined in the conflict, ranging from Japan in the east to the United States in the west. Over the next several months the government in Petrograd (patriotically renamed from Saint Petersburg in August 1914 by replacing the German "burg" with the Russian "grad," for city) mobilized its armed forces against Austria; Serbia was something akin to Russia's godchild. A Slavic country, Serbia shared Russia's Orthodox faith and Cyrillic alphabet; indeed, its very existence emerged from the fallout of the Russian war with the Ottoman Turks, who had controlled the Serbs. When Russia also mobilized against Austria's ally, Germany, the Germans declared war on Russia on August 1, 1914 (N.S.).[1]

At that point, the Imperial Russian Army had access to 244 aircraft. Its naval counterpart flew an additional 29 airplanes; 4 of these were land planes and most of the rest were flying boats with hulls or else seaplanes that were equipped with pontoons. The Russian navy designated all aircraft that took off from and landed on water as *gidroplan* (hydroplanes)—the translation of an American term used by the U.S.-based Curtiss Aeroplane and Motor Company (as it eventually was known), which built many of the flying boats and seaplanes that the Russians acquired. Whether the aircraft were built in Russia or purchased abroad, the Russian Empire seemed to be as well-prepared for air combat as any of the belligerents when the guns of August heralded the start of a terrible conflict in 1914. In fact, Russia's widely varied aircraft fleet outmatched the airplane inventory of Germany, which later emerged as its primary wartime enemy. This may come as a surprise to those who are familiar with Russia's development before the start of the twentieth century. The empire's unbridled monarchy, which claimed absolute powers, was not noted for its progressive attributes.[2]

Because of Russia's love-hate posture regarding Western Europe and the United States, rarely did its emperors imitate the dramatic behavior of Tsar Petr the Great. A member of the Romanov family, he technically ruled the empire starting in 1682, but actually exercised power only between 1689 and 1725. Petr imported Western products, practices, and personnel in his forceful—but not always successful—attempt to modernize his Slavic domain. On the other hand, Tsar Nikolai II, the Romanov who ruled before and during the Great War, was at least open to the kind of Western technology that could be used in military hardware. The reason is not hard to discern. Russia suffered a series of disastrous defeats on both land and sea in the Russo-Japanese War of 1904–1905. It was an extraordinary moment in world history when an island country of Asia whipped the world's largest empire. Small wonder that Nikolai Romanov and his semi-autocratic administration suddenly recognized that the empire's military needed serious modernizing, from artillery to warships and everything in between. After 1905 Russia spent millions of rubles on war materiel; unfortunately, its military leadership remained backward throughout the empire.[3]

Meanwhile, by May 1905 the victorious Japan had almost exhausted its resources—both men and money—after having pummeled Russia's army in Manchuria and sunk most of Russia's Baltic Fleet to the bottom of the Sea of Japan. The shortcomings prompted the government in Tokyo to ask President Theodore Roosevelt secretly to broker a peace settlement. He later was awarded the Nobel Peace Prize for bringing the warring countries together at Portsmouth, New Hampshire. At a meeting there in September 1905, the Russian and Japanese delegations agreed to sign the Treaty of Portsmouth. That same month the Wright brothers, Orville and Wilbur, offered demonstration rides in the Wright Flyer—an early model airplane that Tom D. Crouch, senior curator of the National Air and Space Museum, later would call "one of the most extraordinary machines in the history of technology." Crouch's assessment clearly was not based on the few piloted test hops that had been flown in the Wright Flyer aircraft in December 1903. The Wrights were not the first to carry potential aviation devotees. Several prominent aviators from the United States and Europe already had taken hops in aircraft—among them Frenchmen Félix du Temple de la Croix (1874) and Clément Ader (1890) and American Augustus M. Herring (1898). As early as 1896 Samuel Pierpont Langley, secretary of the Smithsonian Institution, assisted by inventor Alexander Graham Bell, launched the world's first successful powered (though unmanned) airplane, which flew 84 meters (about 275 feet).[4]

Of special interest was a monoplane designed by Russian Aleksandr F. Mozhaiskii that "flew" in 1884. Son of an admiral in the Imperial Russian Navy, Mozhaiskii was born in southern Finland (then part of the Russian empire) in 1825. He followed his father in naval service; his time in the military was interrupted, but he nevertheless retired as a rear admiral (*kontr-admiral*) in 1882. An important development in his later work was designing a motorized heavier-than-air craft sometime in the 1850s, when naval duty took him to Japan. As the island country rapidly surrendered its isolation, Mozhaiskii demonstrated a working model of a steam engine, which the Japanese carefully copied. Understandably, when he first drafted plans for an airplane, as a naval officer, he already knew what the craft's motor would be. In fact, two steam engines propelled his monoplane, which reportedly traveled in the air for about 25 meters (82 feet) in a hop in 1844. The heavy, underpowered, and flat-winged plane was incapable of providing much lift; it managed a short flight mainly because it had rolled down a ramp to gain momentum. After decades of high Soviet praise for the 1884 flight, even the Soviet Union's Central Aero-Hydrodynamic Institute (TsAGI) finally had to admit that Mozhaiskii's brief hop had failed to accomplish true powered flight.[5]

By contrast, during September and October 1905 the third version of the Wright Flyer proved to be truly the world's first genuine airplane. It flew for a respectable distance under complete control, changing its attitude, turning and returning to the field where the flight had begun. Thus the year 1905, not 1903, marked the real birth of human flight in heavier-than-air craft. Even so, Russian aviation seemed to catch up to America's—and even surpass it—after 1905. By 1914 Russia's military was able to maintain much better and far more aircraft. Most of the fifteen aircraft procured by the U.S. Army and Navy before the Great War had to be discarded as obsolescent and unsafe. Manufactured by the Wright Company, they were some of the worst-built airplanes in the world—slow, tail-heavy, unstable, and derided as "man-killers" by military pilots. Moreover, the Wright Model C continued to use awkward and confusing double levers to alter control surfaces rather than the single stick and rudder pedal found in European craft. The Wright Company also failed to use dihedral wings, an omission that made the plane tricky to fly and forced pilots to manipulate the twin levers constantly to enable the Model C to maintain a safe attitude during flight. As a result of these deficiencies crashes occurred and several U.S. Army pilots died before the Wright aircraft were grounded in 1913 and then scrapped.[6]

Meanwhile, the Russo-Japanese War of a decade earlier revealed to the Russian military the benefits of human flight that preceded airplanes. During that conflict the Russians used manned balloons for observing and gathering intelligence on fixed Japanese positions in Manchuria. The success of balloon reconnaissance in combat had two results. First, the number of balloon detachments grew after 1906 and laid a foundation for the future use of aircraft in a reconnaissance role. Second, the Imperial Russian Army entered the world of flight in the mid-1880s by establishing a training park for aeronauts at Volkhov Field, east of Saint Petersburg. The Russian action came slightly more than a century after three Frenchmen invented the basic two types of balloons in 1783; heat lifted the lighter-than-air craft built by Joseph and Jacques Mongolfier; hydrogen gas did the same for a balloon designed by Jacques Charles. The lifting agent for the latter balloon became the standard for most armies and navies. The very first military air unit was authorized in Revolutionary France by the governing Committee on Public Safety on April 2, 1794.[7]

Because it took decades for the Russian military to create its first balloon unit, it is easy to suggest that the time lag fit neatly into the pattern of Russian backwardness. The U.S. Army flew balloons during the Civil War that began in 1861, officially named the War of the Rebellion. After watching citizens in hydrogen flight after 1804 the Imperial Russian Army certainly realized that balloon flight presented interesting problems in combat. A free-floating hydrogen balloon can change its altitude by dropping ballast or releasing some of its gas, but the flight can only move in the direction of the prevalent wind. If a military balloon were blown over an enemy position, it could not turn around and return to its starting point. As a result, such free flights could easily qualify as suicide missions. For this reason, the pan-European 1899 Hague Conference on arms reduction rejected a Russian proposal to ban dropping explosive projectiles from balloons. To the Russians, the most sensible wartime mission for a manned balloon was to gather information on an opponent through a tethered flight a mile or more out of range from the enemy's fixed position. Indeed, in their victorious conflict with the Ottoman Empire in 1877–1878, the Russians were forced to conduct an extremely bloody assault and lengthy siege at the Turkish fortress of Plevna (Pleven in modern Bulgaria) that lasted for several months. The fact that the troops at that battle were forced to remain stationary during the conflict was part of the reason that the Russian military took an interest in lighter-than-air craft a few years later—and created the Aeronautical Training Park in 1885.[8]

Battles associated with the Russo-Japanese War of 1904–1905 only confirmed the value of balloons as an integral part of army reconnaissance. By 1907 the empire possessed several field and fortress units. Five years later it boasted 13 balloon detachments. Varied tasks prompted detachments to have flexible numbers of 1 to 4 balloons with a pilot-observer for each, several officers, and ground support ranging from around 35 to 195 soldiers. After the Great War began, Tsar Nikolai II ordered that the bulk of the balloons be based in Vladivostok. The emperor wanted to protect the

Far East city because it was only one of two Russian ports still open to seaborne traffic and international trade. During the war, the Aeronautical Training Park doubled its military personnel to 14 officers and 216 troopers. The park included workshops, aeronautical equipment, photographic laboratory, and a pigeon-training station. In the absence of radios, birds provided a means of communication with ground forces.[9]

The Plevna siege also strengthened Russian resolve to maintain and enhance a line of fortresses along western portions of the empire, in the former territories of Poland that had been absorbed by Russia near the end of the eighteenth century. The fortresses at Kovno, Grodno, Osowiec, Novogeorgievsk, Warsaw, and Ivangorod held troops, along with thousands of high-trajectory, large-caliber guns and millions of shells for cannons, rifles, and machine guns. All the attention, troops, treasure, and hardware provided to the citadels were designed to make them impregnable in the face of potential German forces bent on penetrating the lands of European Russia. Precisely because these strongholds epitomized fixed positions, they were given balloon detachments. Yet, the truth was that the citadels were useless and a sad waste of war materiel. Not just in Russia, but in Western Europe as well the fortresses at Liège, Namur, Maubeuge, and Antwerp ended life as fortified traps that had to be evacuated or surrendered after falling victim to German heavy artillery. Since the Russian military leadership regarded the fortresses as sacrosanct, field units would suffer especially in 1915 because they lacked the artillery or shells that were being shipped to—and held by—these dinosaur-like bastions.[10]

While balloons and fortresses initially helped define the important fixed positions on the Eastern Front, the same could not be said about the Western Front, in which Anglo-French troops were pitted against their German counterparts. For four hundred miles, from neutral Switzerland to the English Channel, each side dug an elaborate trench system that was the site of horrific blood and death for the next four years. Unlike the situation on the Eastern Front, the entire trench system in the west served as a fixed position, with only minor changes from time to time. Naturally, both sides on the Western Front launched—and lost to aircraft—a large number of balloons in the effort to observe the enemy and spot artillery. By contrast, although the Eastern Front was more than twice as long and just as bloody as the Western Front, the war in the east remained more fluid, especially in 1914 and 1915 and in the southern section in 1916. Only portions of the front in 1916 opened the possibility of using balloons on a regular basis. By then, however, aircraft carried machine guns, manned by enemy pilots who considered balloons prime targets for destruction.[11]

It would be unfair not to mention that Russian pilots reciprocated. Boris Sergievsky, for example, flew a French Nieuport fighter built under license by Russia's largest aircraft manufacturer, the Moscow-based Dukh Company. Late in the war Sergievsky specialized briefly in shooting down balloons. He approached them at an altitude of 16,000 feet—fairly safe from exploding antiaircraft shells fired by the Germans. When a shell burst below his plane, he would fake being hit and drop straight down in a tailspin. Thinking the Russian plane had been fatally hit, the Germans stopped firing their guns. As he flew close to the balloon, Sergievsky would straighten the plane and fire incendiary bullets into the side of the hydrogen-filled balloon. The gas would burst into flames as he quickly dove, pulled out, and then flew parallel to the ground. He then skimmed across German lines too fast for soldiers to see him in time to riddle him and the Nieuport with rifle or machine-gun bullets. After an exciting but failed effort to destroy another balloon at the same site, Sergievsky quickly learned that Germans were smart enough not to fall for the same trick a second time. From then on, he never returned to a location where he had destroyed a balloon by trickery.[12]

For Russians the obvious vulnerability and static nature of balloons heightened the importance of aircraft for reconnaissance and diminished the value of lighter-than-air craft—especially when the war progressed under fluid conditions. With that in mind, in the period before 1914 they experimented with an early version of the dirigible that seemed to show great promise. The invention comprised a tubular envelope that enclosed one or more hydrogen-filled balloons and included some means of navigating the craft in flight. Actual, full-scale dirigibles from experimenters such as Pierre Jullien and Jules-Henri Giffard first appeared in France in 1850 and 1852. The craft were powered by clockworks in one case and by steam engine in another; both turned propellers that would move each dirigible through the atmosphere, and they gained or lost altitude by shifting the vehicle's center of gravity. The capstone period for the evolution and emergence of large, fully navigable lighter-than-air craft occurred between 1900 and 1908, under the direction of Graf Ferdinand von Zeppelin; his name ultimately became a synonym for dirigible. The final product had a rigid frame and internal combustion engines that rotated propellers; it also had control surfaces including multiple elevators, horizontal stabilizers, and vertical rudders.[13]

Development of the Zeppelin did not occur under the auspices of the German army or government, and it most assuredly was not a secret project. Public campaigns to raise funds for the new technology attracted donations from citizens across the country. In fact, successful tests of the gigantic dirigibles gained international attention. Since the craft had the obvious potential for serving in a bombing or scouting role for the military, the Russian government could not ignore what was happening in Germany. In 1907 it created a Commission for the Planning and Construction of Piloted Aerostats. The group began experimenting and in 1908 completed a prototype for a non-rigid training dirigible. The following year the Imperial Russian Army acquired its first version of the craft, the *Krechet* (large falcon). Eventually, the Deka Company of Aleksandrov started building the dirigibles for the military. On January 1, 1914, Russia's dirigible force took fourth place worldwide—behind Great Britain, Austria-Hungary, and Japan. Russia had six large fortress dirigibles (*Krechet, Grif, Al'batros, Kondor, Burevestnik*, and *Astra*) and nine smaller field dirigibles (*Lebed', Berkut, Iastreb, Golub', Sokol, Korshun, Chaika, Kobchik*, and *Mikst*). By August 1914 the field dirigibles were out of date and were incapable of flying faster than 35–55 kmh (about 23–34 mph)—an insignificant speed. Only three dirigibles, *Kondor, Berkut*, and *Al'batros*, took part in battle action. Unfortunately, the slow craft required special maintenance, carried flammable gas, used scarce engines, experienced fiery accidents, and suffered hazard from enemy guns. By the end of the first year of the Great War, Russia had grounded its dirigibles as unsuitable for any type of combat role.[14]

The rapid downfall of the military dirigible cannot be fully understood without recognizing how quickly the airplane took over every possible combat mission that might be considered for, or assigned to, a navigable lighter-than-air craft. In Russia, the initial upsurge of interest in heavier-than-air craft followed the well-publicized flight of 722 feet in France by a diminutive Brazilian, Alberto Santos-Dumont, in a primitive aircraft (No. 14-bis) on November 12, 1906. (Indeed, everyone thought that Santos-Dumont had developed the world's first real airplane; the Wright Brothers remained secretive until 1908, when Wilbur flew demonstration flights in Europe while Orville did the same for the U.S. Army.) Meanwhile, the Russians, who lacked a reasonably powerful but light engine, had turned to building gliders. Early in 1908, along with some balloonists, these glider enthusiasts formed the Imperial All-Russian Aero Club. Their hope centered on creating aircraft by merging a few glider projects with imported motorcycle or motorboat engines from the West, such as the "Antoinette" that Santos-Dumont used.[15]

French engineer Léon Levavasseur designed the motorboat engine and named it for Antoinette Gastambide, the daughter of his employer. As one can imagine, the Antoinette engine enjoyed great popularity among pioneering pilots and airplane builders in France between 1907 and 1909. The country, in fact, became famous for its engines from several different manufacturers. More often than not, Russian aircraft housed French-designed internal combustion engines during the Great War. One aircraft engine, built by Italian motorcycle racer Alessandro Anzani, played a key role in an event that jump-started the acquisition of a Russian air force. An Anzani motor powered French aviator M. Louis Blériot's Model XI monoplane aircraft from France to England when he flew across the English Channel on July 25, 1909. To be sure, it was only twenty-three miles from Calais to Dover, but Blériot's flight gave him the same kind of international fame that later was accorded Charles Lindbergh for his 1927 solo flight across the Atlantic Ocean from New York to Paris. At a subsequent luncheon held by Blériot, Lindbergh admitted to his host: "I shall always regard you as my master."[16]

The global press accounts that followed Blériot's 1909 flight turned the event into a decisive moment for the Russian military's interest in heavier-than-air craft. As it turned out, Grand Duke Aleksandr Mikhailovich Romanov, the cousin and brother-in-law of Tsar Nikolai II, happened to be in France at Biarritz on holiday when he read newspaper accounts of Blériot's epochal flight. The feat made clear to Grand Duke Aleksandr that the airplane could transcend natural and fortified barriers between countries.

Even in the reconnaissance role to which the world's military first assigned them, aircraft had the potential of compromising the security of otherwise powerful nations. The grand duke went back to Russia enthusiastic about introducing heavier-than-air flying machines to the armed forces. And he had the resources and dynastic connections to fulfill his wish. Unlike most of his Romanov brothers, uncles, and male cousins, who entered the army as officers in one of the elite Guard Regiments, Aleksandr had graduated from the Imperial Russian Naval Academy and worked his way up from *michman* (warrant officer, but equivalent to ensign) to rear admiral before leaving the navy after twenty-four years' service.[17]

During the Russo-Japanese War the grand duke commanded a squadron of twelve torpedo boats and formed and chaired a committee of volunteers that raised two million rubles from the general population to build additional torpedo boats for the Imperial Russian Navy. The war ended before all the money could be spent, but Aleksandr gained permission from donors to spend 900,000 rubles (about 450,000 U.S. dollars) on airplanes and secured approval from the emperor and the naval and army ministers to send a select group of officers to France for flight training. He purchased aircraft from Blériot and arranged for the training from him and the brothers Charles and Gabriel Voisin, also major aircraft producers. The tsar also approved Aleksandr's plan to reorganize his volunteer group into the Committee for Strengthening the Air Fleet. The group subsequently raised another 226,923 rubles. Small wonder that Aleksandr is viewed as the secular patron saint of Russian military aviation. The grand duke went well beyond those feats, however. Already the owner of an estate located on the Crimean Peninsula, which jutted into the Black Sea, he also purchased land near Sevastopol', an area that enjoyed moderate temperatures suitable for year-round flying, and laid the foundation for a military aerodrome. Both the army and the navy eventually built and maintained technical schools of aviation and flight-training facilities.

He also worked hand-in-glove with the Imperial All-Russian Aero Club, several of whose members temporarily conducted basic flight training for those military officers who did not go to France to become pilots.[18] The Aero Club had a dramatic impact on Russian military aviation. Numerous members ended up working for, flying with, or supplying aircraft to the army or navy during the war, including the Seversky family. Popular singer Nicholas G. Seversky (an anglicized stage name for Nikolai G. Prokof'ev-Severskii) flew a bomber during the war while his younger son, George, trained army pilots. Seversky's older son, Alexander, flew fifty-seven combat missions and shot down thirteen German planes for the Imperial Russian Navy. Another club member, Vladimir A. Lebedev, traveled to France in 1909 and studied with renowned aviation pioneer Henry Farman. The Russian returned home with a Farman Model III biplane. Soon the Dukh Company, then a Moscow-based motorcycle firm, acquired a license and manufactured a large number of Farmans that became the Russian military's primary trainer during much of the war. Headed by Iurii A. Meller, the company later mass-produced French fighter aircraft. Before long, Lebedev formed an airplane firm near the Komendantskii Aerodrome in Saint Petersburg. By April 1914 he had enlarged and reorganized his joint-stock firm into the Lebedev Aeronautics Company. Like Dukh, it became one of the Russian military's major wartime suppliers of license-built French aircraft, including Deperdussin, Morane-Saulnier, Nieuport, and Voisin, as well as flying boats originally designed by Franco-British Aviation.[19]

Meanwhile, the famous Blériot flight of 1909 prompted Nicholas Seversky and other members of the Imperial All-Russian Aero Club to arrange for the manufacture of the Model XI airplane and the import of Anzani motors. The club formed the First Russian Aerostatics Company, which was housed in the Shchetinin Works in the empire's capital city. Engineer Nikolai V. Rebikov used photographs of the Model XI from a correspondent in France to design and construct the Blériot look-alike. Although the Russian copy did not measure up to the original, its only visible difference from the Model XI was the use of a skid in place of the swiveled tail wheel. Christened the *Russia* in 1910, this first airplane from Russia's first aircraft factory was purchased and flown by Seversky himself. Replicated, it became another early trainer for the Russian military. Later, the plant built other French planes, such as Nieuports and Farmans. A year before the war, Dimitrii P. Grigorovich, graduate of the Kiev Polytechnic Institute, became Shchetinin's chief engineer. He began designing and building flying boats—airplanes that had ship-style hulls. His *morskoi* (naval) aircraft, especially the M-5 and M-9, were built by the hundreds and were used effectively by the Imperial Russian Navy in the Baltic and Black Seas theaters. Grigorovich was the most significant aviation engineer of the tsarist period who continued to work successfully as an aircraft builder in post-revolutionary Soviet Russia.[20]

Like Grigorovich, Igor I. Sikorsky attended the Kiev Polytechnic Institute, following three years of studying math and engineering at the Imperial Russian Naval Academy in Saint Petersburg. Sikorsky became the most famous member of the National Aero Club and internationally remains Russia's best-known pioneer in flying and designing aircraft. His interest in aviation was sparked by reading the science fiction of Jules Verne and newspaper reports on Wilbur Wright's flight demonstrations in Europe. Fortunately for Sikorsky, the son of a pre-Freudian psychiatrist, his family had enough resources to fund his pursuit of flight for several years. In 1909 Igor spent time in Paris studying aeronautical principles with French pioneers such as Blériot. He returned to his home in Kiev with an Anzani 25-hp engine and made two unsuccessful attempts at building a helicopter before creating the first of his Winged-S airplanes. In 1912 Sikorsky's efforts were rewarded financially when his S-6 model won the Russian military competition. This triumph led to his employment by the newly formed aviation section in Saint Petersburg of the Riga-based Russo-Baltic [Railroad] Wagon Company. In 1912 and 1913 the firm subsidized his development of a four-engine behemoth, the *Grand*. Its successor, the Il'ia Muromets, would become the world's first four-engine reconnaissance-bomber. Under Sikorsky's direction, and with various revisions, the Russo-Baltic Wagon Company built seventy-five versions of the plane for the empire's use during the Great War.[21]

The fifth major aircraft manufacturer during the war was associated with the Odessa Aero Club, sponsored by the Imperial All-Russian Aero Club. Entrepreneur and banker Artur Antanovich Anatra worked through the Odessa group, which maintained a shop that built aircraft. In 1913 Anatra gained a contract from the War Ministry to construct five Farman IV primary trainers. Soon his Anatra Aircraft Company began building other Farman models, along with Nieuports, Voisins, and Morane-Saulniers. The firm developed at least ten other in-house models, but only two entered into production, both at modest levels. One was a modified Voisin, and the other proved difficult to control in flight. Lebedev and his company also built several indigenous planes, but only one airplane, the Lebed XII (Swan), attracted many military purchases. Unfortunately, the plane suffered from uneven workmanship in serial construction, and the exhaust system routed deadly fumes into the second cockpit, which was reserved for the observer, who also manned a machine gun. Additionally, the aircraft sometimes erupted in fire—an unlucky experience for the pilot in the first cockpit, who sat directly on top of the gasoline tank.[22]

Finally, Aero Club member Sikorsky designed a dozen smaller aircraft built by the Russo-Baltic Wagon Company. Most of these never left the prototype stage except for the Sikorsky S-16, of which the company made 34 copies. Drafted in the fall of 1914, the first S-16 appeared in the summer of 1915. Powered initially by an 80-hp Gnome engine, the airplane served as a pilot trainer for the Il'ia Muromets. After Georgii I. Lavrov created a synchronized machine gun for the aircraft, three S-16s served briefly as fighters in the Seventh Army Detachment in 1916, led by future ace pilot Ivan A. Orlov in the spring. But the slow plane suffered construction problems, could not reach higher altitudes, and lacked replacement supplies, and the synchronized machine gun only fired into propeller blades at certain speeds. Sikorsky quickly returned the S-16 to its original task of training pilots how to fly.[23]

Obviously aero clubs and their members played a part in forming the manufacturing infrastructure for military aircraft. In wartime 14 other workshops built or tried to build airplanes. The 5 firms mentioned above—Anatra, Dukh, Lebedev, Russo-Baltic, and Shchetinin—turned out most of the approximately 5,550 aircraft that Russia manufactured while it participated in the conflict. Except for Russo-Baltic, the companies began as modest workshops before 1914 and grew into larger facilities. By 1916 they employed hundreds of workers, who built the aircraft out of wood and fabric, 1 or 2 a day. At the same time, the other 14 aircraft shops collectively produced 6 to 8 planes a month. Even so, the majority of airplanes built in Russia were copies or slightly modified copies of French aircraft, and most of the 900 or so imported planes were built in France or, sometimes, Great Britain. The imports always represented models from the previous year. The United States exported only a few naval aircraft before and during the war; fortunately, they were not built by the Wright Company.[24]

Later, when the Soviets listed Russian inventions and claimed to be the first in the world to create everything from airplanes to zippers, aviation historian Vadim Borisovich Shavrov argued that Russia had matched the capitalist powers in aviation development during the course of World War I. He failed, though, to highlight that France produced far more aircraft than Russia in one year (1918), or that one U.S. manufacturer, Curtiss Aeroplane and Motor Company, built almost as many planes as Russia after America finally declared war on Germany in the spring of 1917. Instead, Shavrov focused on the creation of 280 experimental aircraft and claimed that 38 Russian-designed planes had entered series production. Although Shavrov made a good point about widespread experimentation, his comment about original Russian aircraft must be qualified. He included aircraft that were merely a step or two beyond the first prototype and those that had been only slightly modified from previous models, as well as planes found unacceptable for combat operations. As suggested earlier, worthy and serviceable Russian aircraft emerged mainly from the work of Grigorovich and Sikorsky.[25]

Nevertheless, the origin of Russian airframes was not a major issue in wartime. Aside from original seaplanes and flying boats, the only land-based warplane built in America and actually used in Europe during the Great War was the D.H. 4, a single-engine biplane reconnaissance-bomber. Its airframe was designed in Great Britain by Geoffrey de Havilland, but the airplane was merely a limited glider, without a decent motor. The U.S. version of the D.H. 4 was powered ultimately by the famous Liberty, a twelve-cylinder liquid-cooled engine that generated 410 hp. Two automotive designers conceived the motor, which was produced by the thousands in engine shops belonging to Packard, Ford, Lincoln, Buick, Cadillac, and Marmon. Some have described the Liberty as America's greatest technological contribution to World War I. By contrast, Russian engines proved to be the Achilles' heel of its military aircraft. Before the war the empire had imported its aero engines from France and Germany. Sikorsky's *Grand* and his first Il'ia Muromets, for example, were propelled by Argus engines built in Germany.[26] (Il'ia Muromets was the model name given to the first aircraft manufactured using the design pioneered in the *Grand*.)

Unlike the design of some indigenous Russian airframes, internal combustion engines demonstrated all too clearly one aspect of the empire's backwardness: Russia simply lacked the experience, technology, skilled workers, and precision machine tools to manufacture certain types of complex equipment. The empire could build a railroad car, but an internal combustion engine posed a difficult challenge. In 1914 there were two basic types of gasoline-powered aero motors—liquid cooled and air cooled; the air-cooled engine was the most prevalent, especially for smaller aircraft. A complicating matter for the Russians was that the common air-cooled engines of the time seemed as intricate as a Swiss watch. The cylinders and propeller rotated around the crankshaft. Such a rotating engine required incredible perfection in construction and balance. It also promoted a torque problem that became stronger as horsepower increased after World War I. France, which had been Russia's military ally since 1894, provided a partial solution to the empire's aero engine dilemma. Motor manufacturers Gnome and Le Rhône opened subsidiaries in Moscow in 1913; they joined together after their companies' merger in France in 1914. The resulting Le Rhône engine, with an initial power rating of 80 hp, was produced in 1914–1915 at the rate of 15 a month. Two years later, with 425 workers, the firm assembled some 40 motors a month.[27]

Subsequently, Salmson and Renault also established subsidiaries in Moscow once the war began. Salmson aircraft engines had power ratings of 130 to 150 hp, and by 1917 the shop could produce almost 50 motors a month. By the March 1917 phase of the Russian Revolution, the several French firms had assembled, often from imported parts, about 1,000 aero engines. Thus some 4,000 additional engines had to be imported—not an easy process because of wartime conditions. The French engines installed in Russian-built aircraft had to be supplemented. The military scavenged motors from enemy aircraft that had been shot down over Russian-held territory or had landed there accidentally. As the war progressed, five Russian companies joined the French subsidiaries in building or trying to build engines to overcome the shortfall in engines. The best of these came from the Russo-Baltic firm, which maintained well-equipped shops and skilled machinists to build decent motors for the Il'ia Muromets. The company had built Russia's first automobile, and by late 1916 it manufactured a Renault 225-hp aero engine. Finally, while engines remained a perpetual shortcoming and caused delays in aircraft deliveries, a key element of flight secured a Russian solution. Several companies, including former furniture makers, fashioned wooden propellers that pushed or pulled planes through the atmosphere. Two Petrograd companies alone, Meltser and Airscrew, crafted a total of 3,795 propellers during the war.[28]

Propellers, along with engines and airframes, represent only the design and physical nature of an airplane; the other important element in aeronautics involves the theory and practice of aircraft operation in flight. Russian aviation enthusiasts, including those in the military, soon realized that a good pilot must understand both the machine itself and its performance in the air. The latter requires an aerodynamicist, and Russia had a great one in Nikolai E. Zhukovskii. Vladimir I. Ulyanov, who led Soviet Russia's first government under the revolutionary pseudonym of Lenin, called him the father of Russian aviation. Zhukovskii earned a doctorate in applied mathematics and joined the faculty of Moscow University as professor of mechanics. As early as 1902 he built the empire's first wind tunnel and used it to calculate the lifting capacity of various wing forms. He also established the Institute of Aerodynamics in 1904 and created an aerodynamics laboratory at Moscow High Technical School in 1910.[29]

At the technical school, Zhukovskii and several of his former students set up a pre-flight aviation program for the military to introduce potential military pilots to the fundamentals of aircraft and the dynamics of flight. Candidates who successfully completed the program transferred normally to the empire's largest and best-equipped flight training center, next to Gatchina Palace, south of Saint Petersburg. The flight center enjoyed its inaugural moment in 1910 when the Imperial All-Russian Aero Club held an air show and then an aviation festival at the Komendantskii Aerodrome in the capital. The attractions drew military officers and hundreds of Russian citizens who, in the words of Grand Duke Aleksandr, "gasped and cheered," in seeing for the first time several different aircraft, including *Russia*, actually flying. One event included a cross-country flight of about twenty-five miles from Komendantskii to the military air park that was being developed adjacent to the palace. Originally the site of an Italian-designed summer home for Empress Ekaterina (Catherine) the Great, who ruled from 1762 to 1796, the vacant lands near the Gatchina Palace had been set aside for military aviation, thanks in part to the encouragement of Grand Duke Aleksandr. Once the center was up and running, it flight-trained army and navy pilots from the Baltic Sea region. Gatchina also prepared other aviation-related personnel, such as observers and mechanics.[30]

Within a few years the military added a dozen more flight centers across the empire. The quality of graduates varied, however; as it turned out, many of the new pilots were woefully unprepared for combat missions. Gatchina remained the exception. During the war, flight candidates spent three months at the Moscow School of Theoretical Aviation headed by Zhukovskii. His lecture notes were later published as *Dinamika polëta* (*The Dynamics of Flight*). Those who passed the ground course took a train to Saint Petersburg (later renamed Petrograd), with a transfer to Gatchina.

Greeted by Gatchina's commander and staff, the candidates received assignments in barracks. For the next several days, they observed operations, learned the lingo of military aviation, and became familiar with the hangars, repair shops, and the officers' club, hospital, parade grounds, aircraft assembly area, and large encampment of conscripted troops. The soldiers provided cleanup crews, provided security for the facility and railroad station, and supplied muscle for moving equipment and turning propellers to start aircraft engines. Flight candidates were not assigned to classes as a group, but were placed over time in a training squadron once a vacancy became available.[31]

Even in the midst of war, flight training at Gatchina took three months and was divided into several segments. Student pilots began their lessons by spending time handling and steering an underpowered Blériot Model XI or another older plane on the ground to gain a feel for the sights, sounds, and controls of an aircraft. The "groundlings" performed their orientation tasks for a number of days on a lengthy strip. It bordered railroad tracks on one side and a stand of trees on the other, which separated the isolated field from the main aerodrome. Once the groundlings became comfortable with the craft, they entered the primary stage of actual flight. The Farman IV served the role as the initial training plane for much of the war. Powered by a Gnome or similar rotary engine rated at 50 hp, the Farman possessed enough capacity to carry two people at the exceptionally tame speed of 35 mph. In one sense it resembled the earliest pusher planes associated with such American pioneers as the Wright Brothers and Glenn H. Curtiss. As in the Wright Model A, the pilot and passenger sat on the leading edge of the lower wing. The Farman, however, did not have multiple controlsticks and wing warping to change the attitude of the plane like the Model A. Instead, a single stick moved ailerons and elevators of the Model IV, and a pivoting crossbar operated by the pilot's feet manipulated the rudder left or right.[32]

After an instructor explained the main points in handling an airplane, he took the student for several orientation flights at the start of the flying phase. Instructor and student sat in the open in a tight tandem configuration that enabled both to have access to the tall stick. During later sessions the student began practicing takeoffs and then landings before attempting a shallow turn. The Farman had such limited power that safe turns required a slight descent to pick up enough speed to avoid a stall. The aircraft stalled when the lifting power of airflow over the wings broke down and turned the airplane into a falling rock. Additionally, the Gnome rotary lacked a carburetor and operated only at full throttle. Speed could be regulated by flipping the ignition switch between on and off. Unfortunately, the Model IV had an eccentric characteristic: turning off the engine in flight caused the tail to drop. This meant that landings had to be accomplished in a powered glide that ended by the pilot's switching off the ignition and leveling the plane just before the wheels touched the ground. Once the student succeeded in solo flight, he moved on to advanced training by flying an assortment of aircraft with more powerful engines, such as the Farman XVI and XX. At Gatchina, during wartime at least, the final hurdle was a written exam which, if passed, resulted in graduation and an officer's commission signed by Tsar Nikolai II.[33]

The fact that advanced pilot training took place through the use of different aircraft hints at an issue not shared by the air arm of the navy. Always a second-class citizen to the million-man army, the navy belatedly explored the use of balloons and even airships before turning to fixed-wing aviation. Besides the significant influence of Grand Duke Aleksandr, an internationally publicized event in the United States nudged the seaborne service in a new direction. On November 4, 1910, Eugene Ely, a pilot for Curtiss Aeroplane and Motor Company, flew a Curtiss Hudson Flier off an eighty-three-foot wooden platform on the bow of a U.S. Navy cruiser, the USS *Birmingham*. The feat inspired the Imperial Russian Naval Ministry to send representatives abroad to evaluate and purchase aircraft. At the same time, a smaller number of sea officers joined army personnel in flight-training programs in France sponsored by Grand Duke Aleksandr and at home by the Imperial All-Russian Aero Club. As August 1914 approached, the Russian navy's inventory of seaplanes consisted almost exclusively of Curtiss hydroplanes and a few land planes. By contrast, the Russian army had at least sixteen different aircraft in service on the date that the war began.[34]

CHAPTER 2

ON THE EVE AND START OF THE GREAT WAR

★ ★ ★ ★ ★

IN NOVEMBER 1910, as the Officers' Aeronautics School was being established at Gatchina, the Sevastopol' Aviation School opened, under the leadership of Grand Duke Aleksandr. His Committee for Strengthening the Air Fleet had an auxiliary group, the Department of the Air Force, which worked with the grand duke in raising money and subsidizing the creation of the southern flight-training center. Because the grand duke had one of the highest possible links to the tsar, there was no question but that the flight-training facility belonged to the military, and not to the grand duke's civilian supporters. Certainly, several commanders of army units joined the festivities of the school's inauguration. On the other hand, the grand duke, himself a retired naval officer, kept a vice admiral as his personal adjutant. As a result, the Imperial Russian Navy maintained a strong presence at the new aeronautical complex. The head of the school's council was a navy captain, and one of the ranking officers to attend the grand opening was the commander of the Black Sea Fleet, Vice Admiral Vladimir G. Sarnavskii. He would later be replaced by Vice Admiral Andrei A. Eberhardt, who served in that post during the war until July 1916.[1]

The school's opening day celebration, attended by high-ranking officers from the navy and army, had a sacramental beginning, with morning ceremonies conducted by a leading clergyman of the Russian Orthodox Church. Armed with cross and holy water, Bishop Alexis, who headed the church in Crimea's administrative city of Simferopol', blessed the school's hangars and planes. The school actually had begun operations two months earlier, with ten aircraft and twenty-nine student pilots, who learned to fly under the direction of the chief flight instructor and pioneer aviator Mikhail Nikolaevich Efimov. After the religious ceremonies, the grand duke and chief instructor took a brief automobile tour of the school's grounds and structures. Events of the day included a photographic session of staff and guests, a late breakfast, and an exchange of telegrams between the grand duke and his cousin, Tsar Nikolai II. Naturally, one highlight of the day's special celebration included flight demonstrations by a couple of instructors who themselves had learned how to handle an airplane at the school in September.[2]

One of the two new instructors, Lieutenant Evgenii V. Rudnev, received several military awards during the war. By the time Russia's part in the conflict ended, with the December 1917 armistice, he wore ribbons and medals of the Order of Saint Anne, Second Class and Third Class, and the Saint George medal. He also had been promoted twice, moving from staff-captain to captain in 1916. These wartime honors explain why Rudnev's military records were preserved, despite the turmoil and disruption of the Russian Revolution and the invasion, bombing, and partial occupation of Soviet Russia in World War II. Conveniently, Rudnev's background typifies Russia's early military pilots. Until the Great War had gotten under way, Russia's traditional social stratification heavily influenced the class of men selected for military pilot training. This was especially true at the Sevastopol' Aviation School. Grand Duke Aleksandr was very much in tune with enlisting young pilots with strong educational backgrounds, who belonged to the state-supported Orthodox Church and whose family held nobility or gentry status and had a history of military service.[3]

Rudnev fulfilled the grand duke's expectations for airplane pilots. His father had served as an officer in the army's cavalry and his family belonged to the Russian Orthodox Church. Moreover, Rudnev had received a strong military education that included aeronautics. As a youth he had attended classes as a member of the Aleksandrovskii Cadet Corps; he graduated from the Nikolaevskii Engineering School as a second lieutenant in 1907. His first few years in the army took an unusual turn. Until 1910 the Imperial Russian Army's Aeronautics Section, which oversaw balloons and airships, fell under the jurisdiction of the Main Engineering Administration of the Imperial Russian Army. Thus, in 1908, after Rudnev was posted to a ballooning company in Vladivostok, army engineers tapped him to attend an officers' course at the Aeronautical Training Park at Volkhov Field near Saint Petersburg. He passed and received a promotion to first lieutenant in September 1909. His first assignment took him to the balloon detachment at Novogeorgievsk Fortress; in late summer of 1910 he was transferred to the Sevastopol' Aviation School. Although buildings and hangars there were still under construction, the school had begun operations.[4]

In 1911 and 1912 Rudnev held various appointments as a pilot-training officer. He wrote two manuals for students. The first presented a discussion of airplane parts; the other described pilot controls, flight operations, and the service requirements for Farman aircraft. In the late winter of 1913 he entered the Emperor Nikolai Academy of the General Staff; he completed the course several months later formally designated as a Military Pilot (*voennyi letchik*)—a title that indicated the bearer's dedication to the tsar, the empire, and military aviation. A month later, seemingly tied to the title, Rudnev received his promotion to staff-captain. The next year, as the Great War began, he trained pilots to fly the Il'ia Muromets reconnaissance-bomber. In fact, he commanded and tested the four-engine behemoth during its first combat mission. (Rudnev had had some experience with disappointing aircraft-testing. The demonstration flight that he conducted for the delight of guests during the opening ceremony for the Sevastopol' Aviation School in 1910 had lasted only two minutes. His engine faltered. Fortunately, Rudnev had had sufficient altitude to lower the plane's nose, gain speed, turn the craft around, avoid a deadly stall, and land the Farman safely with only some fractures to the plane's tail. But his engine problems inadvertently served as a portent of Russia's future difficulties with aviation motors.)[5]

That Rudnev, Alexander Seversky, Viktor A. Khodorovich, and other pilots who attended the Sevastopol' Aviation School or the Officers' Aeronautics School at Gatchina later served with distinction during the Great

War was a testament to the high quality of pilot training offered by those institutions. By the end of 1913, the two schools together had produced ninety-seven pilots. Before the conflict ended, Russia had established other training fields as well—one in Kiev, Odessa, Omsk, Tashkent, Tiflis, and Warsaw (until 1915, when the Polish city fell into German hands) and two in or near Moscow and Saint Petersburg (later Petrograd). At aerodromes other than Gatchina and Sevastopol', problems appeared as the war progressed. Pressure to meet pilot quotas, coupled with aircraft shortages, resulted in accelerated programs. Interestingly, student pilots often received ample ground classes in aeronautics, but then found their actual flight time curtailed. All too often they quickly met with injuries or death in combat missions or airplane accidents. At the same time, the accelerated process and the need for more candidates also tended to diversify the pilot corps: family heritage no longer served as a key factor in the selection of flight school candidates.[6]

Early in their development, the centers at Gatchina and Sevastopol' also took on some distinctive characteristics that were not pleasing to the higher echelons of the officer corps. Gatchina quickly evolved into a type of military academy, with various ground courses on airplane structure and operation and current thinking on how best to use aircraft in combat situations that included strategy and tactics. Flight training at the Officers' Aeronautics School took on new components. Although aviation candidates still focused primarily on learning how to fly, they also began spending time understanding what their role would be in assisting ground forces. Gatchina soon began to train observers and pair them with pilots. Both had to appreciate current thinking on what the army wanted them to accomplish during a typical reconnaissance run. Was the observer to count nearby enemy artillery pieces or supply depots or estimate reserve forces farther to the rear of the enemy's frontline position? Although military aircraft had evolved into fighters and bombers by 1915, scouting strategies continued to be a crucial part of the learning process throughout the Great War. By contrast, Grand Duke Aleksandr and his Department of the Air Force initially had focused most of their attention on basic flight training for both army and navy personnel.[7]

Some military officers questioned the grand duke's enthusiasm for using civilian members of the Imperial All-Russian Aero Club as primary flight instructors for army and navy aviation candidates. A prime example was Aleksandr's selection of Mikhail Efimov to head the flight department at the Sevastopol' Aviation School. A remarkable pioneer aviator, on whom Tsar Nikolai II bestowed the title of Honored Citizen, Efimov never received an officer's commission, which made him the object of denigration by those who had. Such backbiting, along with Efimov's uncertain social status as the grandson of a serf, may partly explain his behavior in the Russian Civil War (1918–1921). He flew for the air arm of the Red Army and was killed in captivity in 1920 by officers of the anti-Soviet White Army.

Grand Duke Aleksandr compounded his problems with the military in other ways. In 1910–1911 he disappointed many in the army and navy who had hoped that Russia would begin manufacturing aircraft on its own; instead, he challenged the anti-West tradition held by many Russians, buying French airplanes after his initial acquisitions in 1910 and continuing to favor that country's aircraft in succeeding years.[8]

In 1912 an outstanding pilot, Lieutenant Viktor Vladimirovich Dybovskii, flew a French-built Nieuport on long-distance flights. An officer at the Sevastopol' Aviation School, Lieutenant Dybovskii fit the standard profile that Aleksandr had sought. He came from a gentry family that belonged to the Russian Orthodox Church, and his father was a lieutenant colonel in the army. Dybovskii flew from Sevastopol' to Kharkov, from Kharkov to Moscow, and from Moscow to Saint Petersburg—a distance of close to 1,500 miles. The grand duke was so impressed by these remarkable flights that he used his close connections with the highest levels of the tsarist government to get the military to buy and distribute Nieuport aircraft among several new squadrons. On August 27, 1913 (N.S.) Staff-Captain Military Pilot Petr N. Nesterov caught the world's attention when he became the first person to fly in a loop—in this case in a Nieuport IV. Alfred and Charles de Nieuport, who started the Nieuport Company, died in separate crashes in 1911 and 1912, but the firm continued and designed and built notable aircraft that were used in World War I by Belgium, France, Great Britain, Italy, the United States, and Russia. Russia built hundreds of them under license. When the first phase of the Russian Revolution began, in late winter of 1917, the army had 545 front-line aircraft; 118 of these were Nieuport models.[9]

The grand duke's success in getting the War Ministry to purchase French Nieuports occurred at the moment that the Sevastopol' Aviation School took the first step in accommodating itself to the broader vision of pilot training that was found at Gatchina. One actor in this process was the war minister, General Vladimir A. Sukhomlinov, whose complicated persona embodied contrasting traits of tradition and reform. On one hand, Sukhomlinov wanted to abolish the fortresses, which he believed were doomed to fail. On the other, he had served for eleven years as head of the time-honored cavalry school—an institution that taught military reconnaissance procedures to horse soldiers. The general laughed aloud in 1909, when he first heard the grand duke's claim that aircraft would replace the horse as the vehicle used by scouts to gather intelligence on an enemy's formation. Sukhomlinov considered aircraft as toys to be played with by wealthy civilians. He clearly placed aircraft in the same category as other newly developed vehicles with internal combustion engines, such as motorboats, motorcycles, and automobiles. Sukhomlinov was not alone. Many other senior officers shared his views on aircraft, especially during the opening months of the Great War.[10]

Fortunately for military aviation—and the near future of Russian airpower—Sukhomlinov experienced an epiphany in 1911. That summer the Imperial Russian Army conducted maneuvers near Kiev and Warsaw. In both cases the Sevastopol' Aviation School supplied aircraft squadrons that conducted scouting missions during the exercises. The planes performed very well, and Grand Duke Aleksandr received high praise for his pilots' work. Military summaries from the two commanding officers in charge of the maneuvers reached Sukhomlinov's desk. They revealed aviation's serious potential in aiding ground forces during actual combat operations. The revelation exposed the side of Sukhomlinov's personality that favored the effort to modernize the Russian army. Realizing that advances in military aviation deserved priority under a system of unified leadership, he issued a decree that transferred the Aeronautics Section of the Main Engineering Administration directly to the General Staff. Next summer the War Ministry created the Aeronautics Unit of the General Staff, to be led by Major General Mikhail V. Shishkevich.[11]

War Minister Sukhomlinov hoped that the General Staff not only would expand aviation as quickly as possible, but that it also would create a degree of uniformity among the pilots and that over time the differences between Gatchina and Sevastopol' would diminish. As early as 1912 novice pilots from Sevastopol' traveled north to Gatchina, entered advanced flight training and classroom courses, and took the examination for Military Pilot. Meanwhile, Gatchina dropped the engineering preference for pilot candidates and, like Sevastopol', broadened its recruiting base. Both had a predilection for candidates who were well-educated, practiced the "correct" religion, and enjoyed a certain type of family heritage. On the other hand, a candidate's military branch no longer was a key issue. By the time the Great War opened, Alexander Seversky's experience in becoming a pilot had demonstrated the similarities and interrelated nature of these two major flight-training centers. After graduating from the Imperial Russian Naval Academy and serving on the torpedo boat *Beaver* during the first couple of months of war, Seversky transferred to the navy's air arm. He studied aeronautics for several months in Moscow, soloed in a Farman Model IV at Gatchina, and then boarded a train for the south. While Seversky took advanced flight training in Sevastopol' in the spring of 1915, he also completed classes on military tactics there. He then entered the navy school and studied naval aviation operations and structural mechanics. Before being assigned to a northern squadron by the Baltic Sea, he was promoted from ensign (*michman*) to junior lieutenant (*mladshii leitenant*). He also completed the requirements and passed the test for designation as a Naval Pilot (*morskoi letchik*), comparable to the army's Military Pilot.[12]

Meanwhile, in September 1912, at the very moment the Aeronautics Unit of the General Staff began operations, the Balkan League considered war with the Ottoman Empire. Bulgaria, Greece, Montenegro, and Serbia had formed a military alliance directed at their former occupier, the Turks. The allies wanted to evict them from Europe, liberate the Albanians and Macedonians, and split liberated territories among league members. Italy's victorious war against the Ottoman Empire to gain Tripolitania (called Libya after 1912) in North Africa provided the catalyst for the Balkan

League's decision to attack the Turks in Europe in October. That same month the headquarters staff of the Bulgarian army sought assistance from Russia's first aircraft manufacturing company, with a request to Sergei S. Shchetinin (who headed the Shchetinin Works, one of the important aircraft manufacturers) for an aviation detachment. In Saint Petersburg (later Petrograd) Shchetinin organized a group of voluntary civilian aviators who called themselves the Special Russian Aviation Squadron. In the third week of October, the group of four pilots and six mechanics traveled to the Balkan War.[13]

The small squadron flew four Farman Model VII aircraft and brought along reserve airplane engines, spare parts, tents, and a fully equipped mobile aviation repair unit, enabling the group to perform reconnaissance duties as a self-contained combat operation. The Russian connection to the Balkan League was palpable because its members had secured their freedom from Turkish rule largely because of Russia's several wars against the Ottoman Empire in the nineteenth century. Due to the ground forces of Bulgaria and Serbia, the league enjoyed a quick victory, followed by a December 1912 armistice with the Turks. Once the fighting ended, squadron pilots, special mechanics, and Shchetinin all received war service decorations from Bulgarian Tsar Leopold I [his full name was Maximilian Karl Leopold Maria]. The battlefield assistance that the squadron provided was followed closely in Russia. Using aircraft to scout enemy troops in the Balkan War seemed to demonstrate the potential value of aviation in any future Russian conflict.[14]

It might be pleasant to say that the war had had a happy ending, at least for the Balkan peoples. This, however, was only the First Balkan War. In a 1904 treaty, Tsar Nikolai II agreed to be the arbitrator in case of a territorial dispute between Serbia and Bulgaria. It made sense, since both countries existed because of Russia and shared the empire's Cyrillic alphabet and Orthodox faith. At the end of June 1913 Bulgaria decided to hurry Nikolai II into an arbitral award by attacking Serbian and Greek troops in the disputed territory of Macedonia. But the Russian tsar rejected Bulgaria's forceful initiative, walked away from his role as arbitrator, and gave his blessing to what became the Second Balkan War. It should come as no surprise that all the Balkan countries, along with the Turks, attacked and defeated Bulgaria in August 1913. As a result, the country lost the territory that it had won in the First Balkan War and failed to reacquire the additional land in the Second Balkan War. As might be expected, the Bulgarians held a grudge against the Russians and the neighboring Balkan countries, especially Serbia. The episode, twelve months before the start of the Great War, directly explains why two years later Bulgaria joined the Central Powers in war with Russia and sent its troops into Serbia to help defeat that member of the Allied Powers and regain the lands that it had lost in the earlier conflict.[15]

For the Russians, the abrupt downturn in relations with Bulgaria had the potential for troublesome consequences. The Imperial Russian Navy's Black Sea Fleet had a close collaborative relationship with the small Bulgarian navy, which comprised only one gunboat and six torpedo boats. Regardless, during the Great War Bulgaria provided Turko-German naval forces with details about Russian operational procedures in the Black Sea and with information about supply depots and battle strategies. Yet, Russia still possessed a dominant fleet in the Black Sea, and the fleet only expanded and grew stronger during the war. Moreover, despite the appearance of a handful of Russian pilots and planes in Bulgaria in 1912, their presence revealed virtually nothing about the air arm for Russia's army or navy in 1915. It was only after the First Balkan War that the work of the Aeronautics Unit of the General Staff began to produce results. The tone of its work was set by War Minister Sukhomlinov. He argued for a goal of forty-five squadrons with twelve planes each; and he expressed hopes that in the future most aircraft would be developed and manufactured by Russians.[16]

In September 1911, to promote domestic aircraft production, the Imperial Russian Army sponsored a military airplane competition at the Korpusnoi Aerodrome in Saint Petersburg. Three aircraft vied for the prize, but only one could carry out all the conditions of the army competition. The winning entry was designed by Jacob M. Hackel, who owned a workshop. Of German extraction, Hackel never became a significant manufacturer of airplanes. His biplane, however, enjoyed success partly because he installed a good Argus engine from Germany, a motor rated at 100 hp. Since only one plane met the army's minimum standards, Hackel did not receive the prize and the contract, but the military purchased his plane for 8,000 rubles

(4,000 dollars). The next year the same event attracted a dozen entrants and, as pointed out earlier, Igor Sikorsky's S-6 biplane won first prize and a contract; the second and third prizes went to the Dukh Company of Moscow, which built copies of two French planes—Farman and Nieuport—under license. The final military competition took place in September 1913. Sikorsky won again with the S-10 and S-11 aircraft; third and fourth place went to French-designed monoplanes—Deperdussin and Morane.[17]

The competitions minimally fulfilled the war minister's desire to use Russian-manufactured planes. Yet, Sikorsky's victories could not hide the fact that the military ended up arming the Imperial Russian Army mainly with French airplanes because of their quality design and ease of serial production. Nevertheless, by the spring of 1913 Russia had made a good start toward the preliminary goal for military aviation that had been set by Sukhomlinov. Within eight months after the Aeronautics Unit began operations, the army had eighteen squadrons with four planes each and seventy-two officer pilots; there were also ten more pilots from the lower ranks. The latter demonstrated another move away from requiring pilot candidates to be members of the nobility or the officer corps. At the same time, 1913 was the year that the Air Force detachment of the Black Sea Fleet Communication Service tested the idea of placing hydroplanes on ships. On August 27 (O.S.), the collier *Dnepr* was equipped with a crane and a Curtiss seaplane #6 on its top deck. The ship embarked from Sevastopol' Bay and anchored fifteen miles from shore. The plane was deposited in the water twice and retrieved both times. High waves, however, prevented the aircraft from actually starting up and taking off between loading cycles.[18]

The navy continued to experiment with loading and unloading Curtiss planes from ships. Early in 1914, using the cruiser *Kagul* as the base, the Russians began to develop a procedure that would load and unload two hydroplanes simultaneously. The system was perfected on February 6 (O.S.), when Curtiss #5, piloted by Lieutenant Viktor Utgof, and Curtiss #7, flown by Ensign Nikolai Ragozin, were lowered to the water at the same time in only a few minutes.

Although Utgof was not quite two years older than Ragozin, he had taught the ensign how to fly the Curtiss. Both (Utgof earlier and Ragozin later) had attended the theoretical aviation course at the Saint Petersburg

MAP 1. Eastern Borders of Germany and Austria with Russia

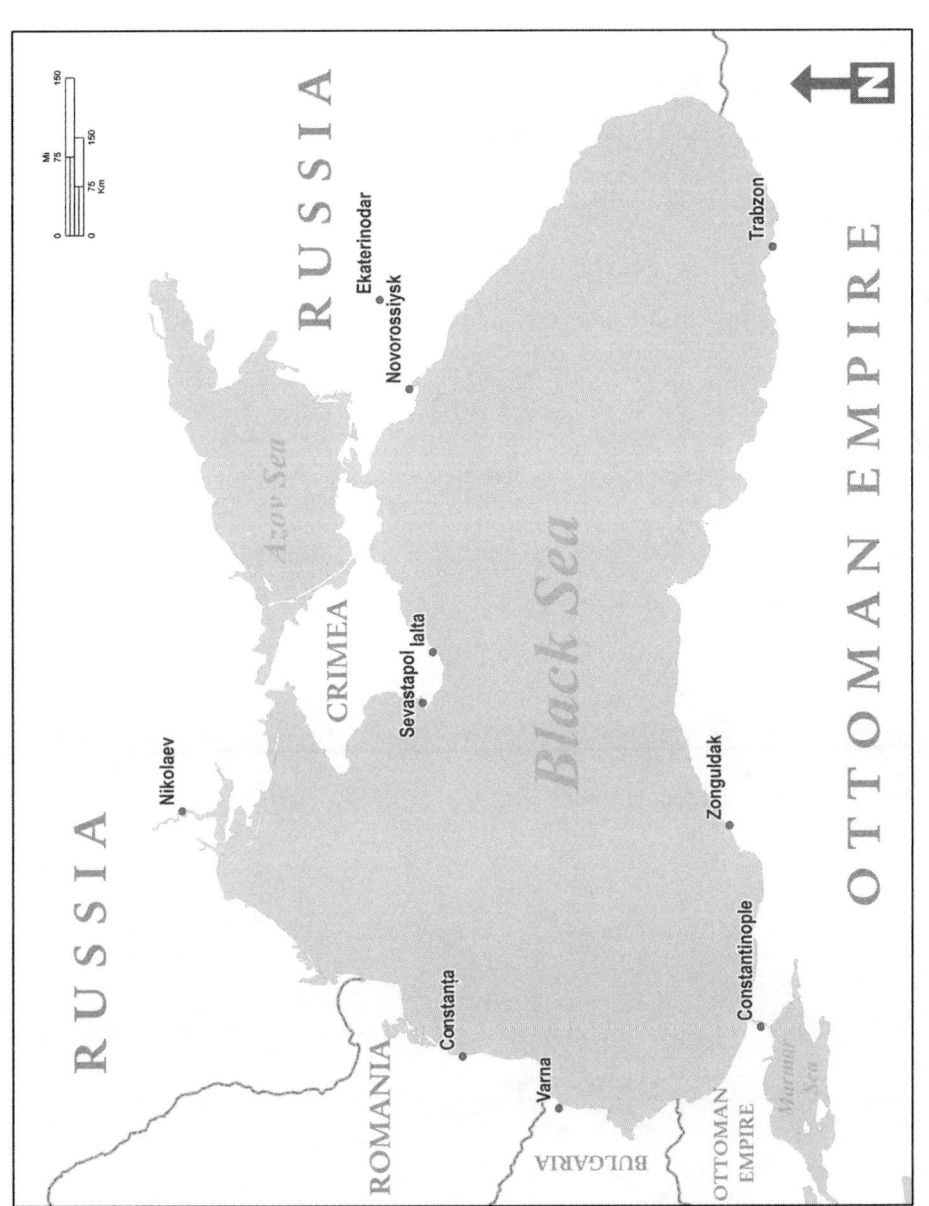

Map 2. Black Sea during the Great War

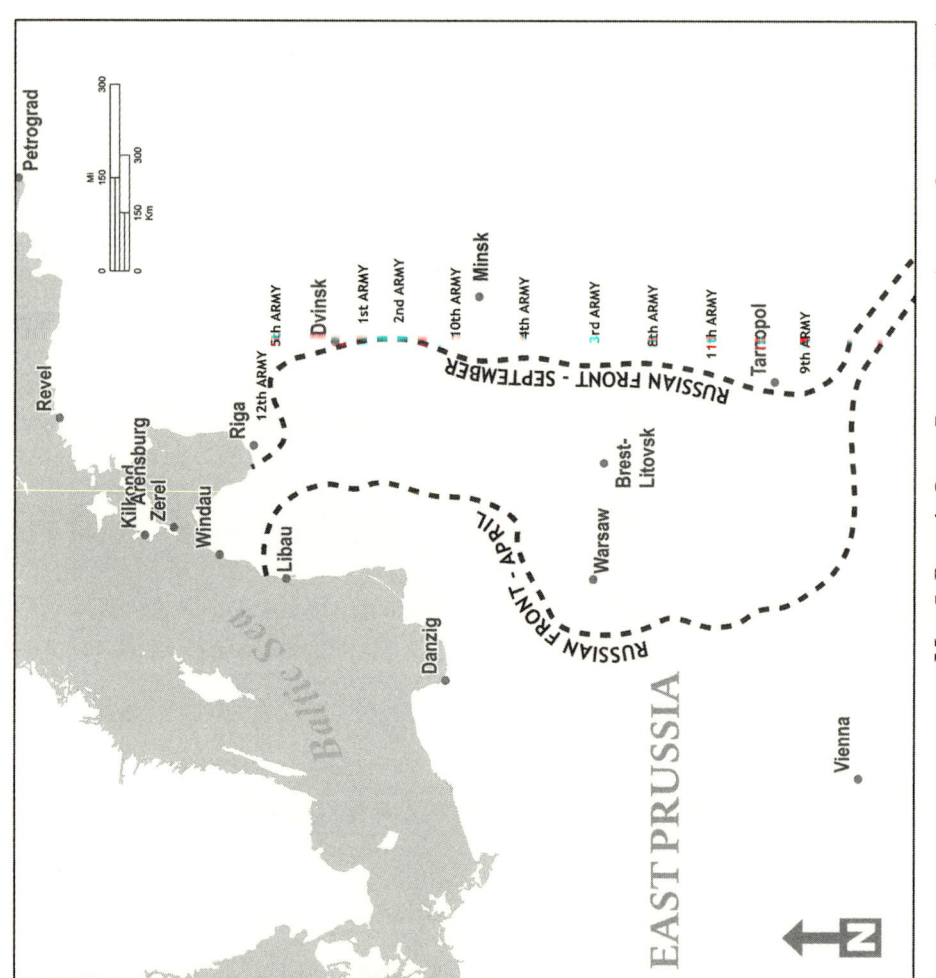

Map 3. Russia's Great Retreat from April to September 1915

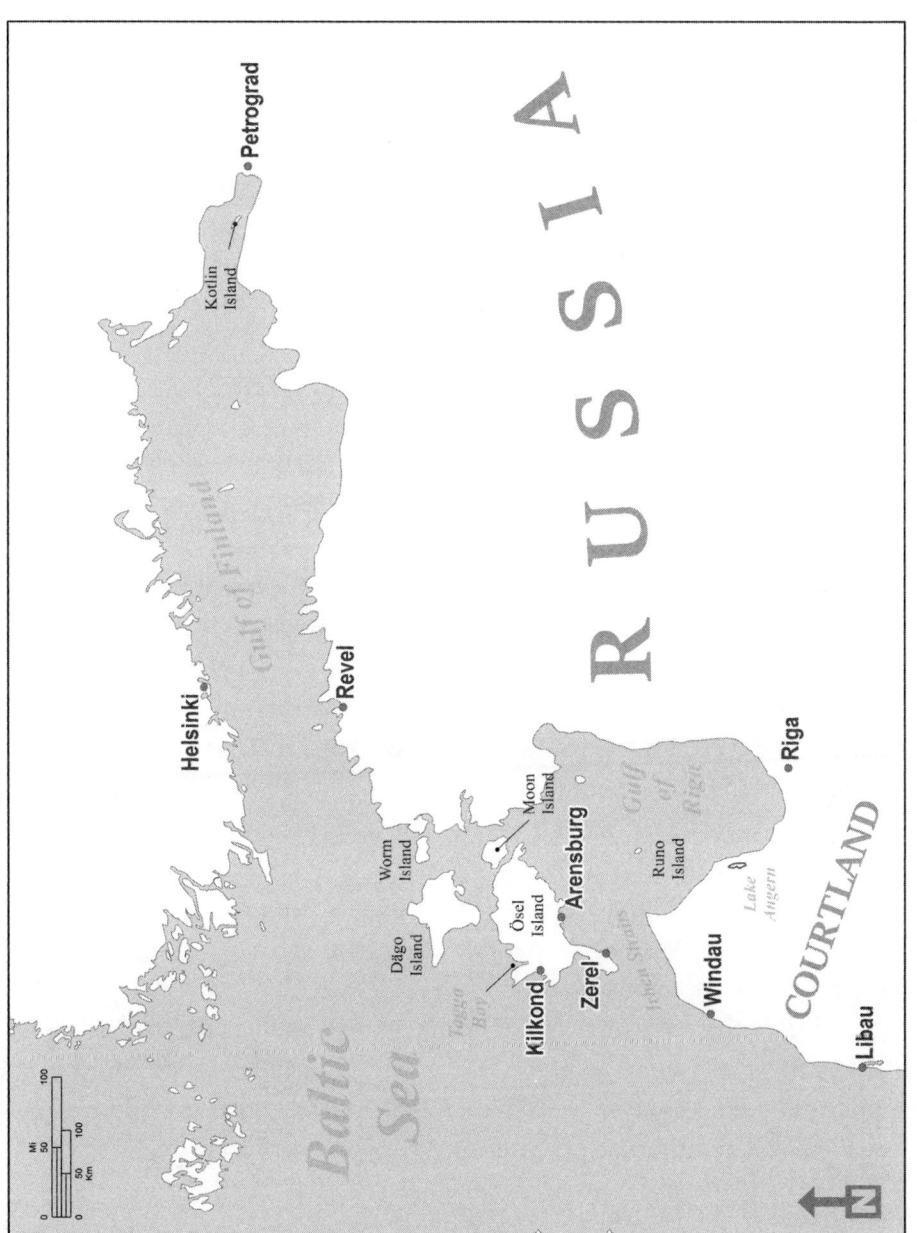

MAP 4. Baltic Sea Region during the Great War

Polytechnic Institute. They also earned awards for outstanding aviation service during the war; and both survived the conflict. Ragozin ended up eventually in Spain; Utgof had been sent as military attaché to the Russian Embassy in Washington. He remained in the United States after the Great War, and in 1923 he helped Sikorsky build his first American airplane.

In late 1914 Utgof, Ragozin, and their planes shipped out on the *Kagul* some sixty miles from Sevastopol'. After their Curtiss hydroplanes were lowered from the deck to the water, they flew back without incident to an air base near Sevastopol'. The air arm of the Black Sea Fleet seemed ready for war.[19]

When the Great War began, the Black Fleet carried six Curtiss seaplanes, seven Curtiss flying boats, two Morane-Saulnier seaplanes, and four landplanes. In the north, the Baltic Fleet only had ten aircraft—Sikorsky seaplanes, Franco-British Aviation (FBA) flying boats, and a couple of land planes. Concurrently, the army had come close to reaching War Minister Sukhomlinov's original goal of forty-five squadrons, although not with twelve airplanes and pilots for each unit. Amazingly, he had projected a standard squadron as having six airplanes for reconnaissance purposes and six additional aircraft to be held in reserve. As it turned out, Sukhomlinov had correctly realized that backup planes would be needed in wartime. His thoughts on the matter back in 1911 had less to do with combat operations than the fact that early aircraft were fragile and susceptible to damage—and even destruction—in normal peacetime use. Regardless, as war approached in 1914 there were forty squadrons, with an average of only six planes each. Six matched the number of aircraft in the squadrons of the French Escadrille and the German Fliegerabteilung.[20]

By then the army air service had undergone another reorganization stage and would soon undergo others. The Aeronautics Unit had been replaced by the Aeronautics Section of the Main Technological Administration, which was actually the renamed successor to the old engineering unit. Regardless, the General Staff retained overall control of aviation squadrons. Similar to the practice in Germany, eight of those squadrons were reserved for the Russian fortresses. The remaining thirty-two were assigned to the thirty-two corps (each with two or more divisions) that made up Russia's eight active armies. Before the war, the squadrons were headquartered at six aviation parks: Saint Petersburg, Warsaw, Kiev, Moscow, Odessa, and Lida in modern-day Belarus. A couple of weeks after war began, the tsar's uncle, Grand Duke Nikolai Nikolaevich, who served as Russia's commander in chief on the battlefield, appointed Grand Duke Aleksandr to head the empire's inspectorate of the military air arm. The new appointee received a railroad salon car for his headquarters at Stavka, then housed in the town of Baranovichi not far from Lida. "The rest," Aleksandr later pointed out, "including airplanes, machine guns, repeating rifles, flyers, observers, technical staff, motor cars, even the typewriters, had to be gotten through my own ingenuity."[21]

Because Aleksandr was married to the tsar's sister, one can question how much of his own "ingenuity" he needed to obtain more typewriters. There is, however, a certain amount of truth implied in the grand duke's description of what his appointment seemed to lack. He ended up on a railroad salon car, in company with those of the General Staff and Grand Duke Nikolai Nikolaevich. The real issue was how much power over military aviation he could actually exercise beyond inspecting the outline of squadron use that already had been put into place by the General Staff. Control of the thirty-two squadrons assigned to the corps and the eight others linked to the fortress system was apportioned neatly: army generals in the field took charge of aircraft units and navy admirals oversaw those in the Baltic and Black Seas. This raised a serious problem, however—at least for the army. Unlike senior naval officers, many field commanders of fortresses and corps demonstrated little or no interest in the potential benefits of aircraft in combat situations.[22]

Corps and fortress commanders had earned the rank of general after witnessing men on horseback reconnoiter "enemy positions" during peacetime field exercises, so it should come as no surprise that some of them, especially those who had not used aircraft in training operations, simply ignored as worthless the intelligence gathered by squadron pilots or observers (who sat in the second cockpit in a two-place aircraft). Yet, after one year of war, planes had proven their usefulness to most field commanders. During the second year of combat, many commanders wanted more aircraft. The situation in 1914 for the Imperial Russian Navy was far different from that of the army. Vice Admiral Nikolai Otto von Essen, who commanded the Baltic Fleet, and his Black Sea counterpart, Vice Admiral Eberhardt, had no men on horseback whom they could send across the seas to pinpoint the whereabouts of enemy ships. For this reason, both admirals prized aircraft intelligence and wanted to increase substantially the number of planes available to their commands.[23]

When the Great War began, the initial course of the action had been designed by a man who had passed away before the conflict actually started. Count Alfred von Schlieffen served as the German chief of General Staff from 1892 to 1906. In his retirement, Schlieffen was promoted from general to field marshal two years before his death in 1913. In his view, mobilization meant war. Thus, when Russia began to mobilize its army against both Austria-Hungary and Germany to protect Serbia at the end of July, the Germans responded with a declaration of war on August 1, 1914 (N.S.). The declaration in essence implemented the Schlieffen Plan that the count had prepared nearly a decade earlier. It was based, of course, on the reality that war with Russia also would mean war with Russia's ally, France. Recognizing the serious military dilemma that Germany would face in a two-front, east-west war, Schlieffen had worked out a scenario to avoid such a calamity. He felt that Russia's backward nature and deficient rail system would force the tsarist empire to take six weeks or more to fully mobilize its forces. As a result, he planned to use those weeks to defeat France first and then attack Russia.[24]

Schlieffen did not want to conduct an offensive on the German-French border, since that area had been heavily fortified on both sides. Instead, he argued for Germany to send several armies through Belgium and invade France in the north, where defenses were less well-developed. The army farthest to the west was the German First Army, under General Alexander von Kluck; it would be responsible for taking the capital city of Paris, which presumably would knock France out of the war. As if on a hinge, five German armies would then rotate and march back to the fatherland, board trains for the rail ride to the Slavic east, and invade and defeat a Russia that had not quite yet finished mobilizing its troops. The Schlieffen Plan promised a short war of movement and a tremendous reward in victory: Germany would then be able to expand its European lands and enlarge its colonial holdings. Two things happened that Schlieffen failed to anticipate. First, the British declared war against Germany because of the Belgian invasion. Great Britain quickly sent troops to aid France, and its huge navy stopped most of Germany's foreign trade. Second, Russia did not need six weeks to prepare some of its forces for military action.[25]

In fact, Russia handed Germany two unexpected blows in August. First, the Baltic Fleet, under Vice Admiral von Essen, initiated a two-month mine offensive. Among other locations, the Russians mined German ports along the south and southeastern shore of the Baltic Sea, ranging from Kolberg in the west to Memel in the east, with other ports such as Danzig in between. The mines sank a cruiser and five transports, disrupting vital wartime trade with Sweden. In the same period the Russian navy brought heavy artillery to fortify the islands at the mouth of the Gulf of Riga. For reconnaissance, the navy also built hangars and other facilities to establish naval air stations for two aircraft squadrons at Kilkond and Zerel on Ösel Island. Later, it housed a third squadron at Arensburg. Every effort was made to prevent German forces from taking the port of Riga, which had a direct rail line to the empire's capital city, Petrograd.

The Russians also managed to acquire a precious gift from the enemy.[26] In the early hours of August 26 (N.S.) the German cruiser *Magdeburg* ran aground in heavy fog near the small island of Odensholm, located in the Baltic about 4.6 miles off the Russian coast. (Today, that portion of the former empire is the country of Estonia.) A lighthouse and a Russian signal station occupied the island. Because of the thick fog, the signalmen could not see the ship, but they could hear German sailors shouting at each other. One of the signalmen radioed the central communications center at Revel about the enemy vessel. Before long the destroyer *Novik*, cruisers *Bogatyr* and *Pallada*, and a half-dozen torpedo boats headed for Odensholm. Meanwhile, the crew of the *Magdeburg* tried to destroy the disabled ship by exploding its magazine, but the blast merely turned the forecastle into scrap metal. An accompanying destroyer took aboard much of the *Magdeburg*'s crew. As the fog began to lift, the Russian ships arrived, forcing the destroyer to make a quick exit. The Russians boarded the cruiser and took its commander and fifty-six crew members as prisoners; the rest of the ship's complement had fled to the island. Almost miraculously, the deck was strewn with papers, among them the signal book and cipher tables, which proved invaluable to both the Russian and the British navies. German naval leaders assumed erroneously that the ship's commander had destroyed these documents and hence preserved their secrets.[27]

Added to the surprisingly audacious naval offensive, less than two weeks after Germany declared war, the Russian army invaded East Prussia with 800,000 soldiers. On August 12 (N.S.) a Russian army division accompanied by cavalry took the town of Marggrabowa, five miles inside enemy territory. Such a quick invasion may have prevented France from suffering utter defeat. It astonished the Germans and drastically altered their war agenda. General Helmuth von Moltke, the German chief of staff, had to order reinforcements for the east, which helped French and British forces stall the German attack in the west during the Battle of the Marne. The stalemate resulted in both sides digging defensive positions to protect troops from machine-gun and artillery fire. It was the start of trench warfare and the end to Moltke's leadership; General Erich von Falkenhayn soon became chief of staff. At the same time the Russians crossed into East Prussia, an even larger Russian force attacked the Austro-Hungarian Empire (Austria). Frantically, the Austrians responded to defend their territory. The Russian attack spared Serbia from confronting an overwhelming Austrian offensive. The small country that in one sense started the war would be able to preserve its autonomy until the fall of 1915.[28]

Aircraft played a small role in the relatively quick start to Russian offensives on sea and land. One example was the squadron assigned to the First Corps. The detachment was headquartered at the Petrograd air park and received orders to transfer to Warsaw. The unit had low priority for travel. It took four days by train for the squadron to arrive at Warsaw; by then the First Corps had already entered enemy territory. The squadron faced immediate problems because its seven-plane inventory included two different Farman models, which complicated maintenance tasks. Moreover, squadron mechanics lacked adequate replacement parts. Once established at Mokotov Field, pilots and support personnel had to change billets almost daily, which reduced unit morale and detachment effectiveness. The treatment of the First Corps Squadron all too clearly revealed the low status of aviation when compared with other elements of the army.

On the Austrian front in Galicia near the town of Zholkov in Lvov Province, the Eleventh Corps Squadron faced a different problem.[29] Petr N. Nesterov, the pilot who gained international fame by being the first to fly a loop, commanded the detachment. He reached a high level of frustration because he had no way of preventing Austrian pilots from flying over the Eleventh Corps and gathering intelligence. Very early in the war pilots on both eastern and western fronts sometimes saluted or waved a hand at enemy planes. Before long such pleasant greetings ended, when the Russians understood that the enemy in the air must be killed to protect "their" men on the ground. Soon opposing pilots exchanged gunfire from pistols and carbines. Pistols, however, were notoriously inaccurate except at very close range, and rifles were not much better because of the movement and shifts in attitude of airplanes in flight. Mounting machine guns on planes or else designing and manufacturing anti-aircraft weapons for use by ground forces came later in the war. As a result, in August Nesterov decided that he would use his airplane as a means of attacking and bringing down an Austrian plane. His decision was not a precursor of the kamikaze attacks by Japanese pilots against U.S. ships and other targets of opportunity late in World War II. While Nesterov had no fear about giving his life for the sake of Russia, he had no intention of committing suicide.[30]

Instead, Nesterov wanted to exercise the *taran* maneuver—a Russian word for a battering-ram. He fully expected to keep his plane in flying condition or at least easily reparable after such action. There were a couple of different ways to carry out the *taran*. For example, he could use the propeller on his plane to chew up the tail structures and control surfaces of the enemy aircraft. They consisted of light wooden stringers covered in fabric. The result might damage or destroy elevators and rudder, and if the tailplane or horizontal stabilizer experienced enough mutilation, the aircraft would take a powered dive to the ground. Another option would be a hard landing of his craft's wheels on the top wing of the enemy plane. It could cause the airfoils, linked together by wires and struts, to fail, causing the plane to collapse toward the ground. The first approach would likely distort or break up the propeller, but Nesterov could turn off the engine and continue flying and then landing the craft as a glider. In theory, the second approach could destroy the enemy plane while preserving a damage-free aircraft for the attacker.[31]

On August 25 (O.S.) an Austrian pilot and observer added insult to injury by flying over and dropping a small bomb on the squadron's field after scouting the position of the Eleventh Corps. Next day when the same Albatros biplane fighter approached the field, Nesterov took off in his Morane-Saulnier monoplane aircraft. At altitude, Nesterov struck his landing gear into the top wing of the Albatros. Both craft began falling to the ground, since the Morane-Saulnier had locked its wheels to the wing of the Albatros. When the Russian plane separated, it went into a spin. Nesterov did have a seat belt, but he did not have it buckled. The spinning craft thrust him out and he fell to his death. Both the Austrian pilot, Franz Malina, and his observer, Friedrich Rosenthal, perished in the crashed landing of the Albatros. The Russian press treated Nesterov as a hero. Tsar Nikolai II awarded him posthumously the coveted Order of Saint George. The pilot was buried in the medieval tomb of a Kievan prince, and he was hailed as a martyr in the struggle to protect Mother Russia. Nesterov's burial in Kiev, Russia's ancient capital, reflected a powerful symbol of patriotism. The heroic treatment of Nesterov encouraged a number of Soviet pilots to use the *taran* against German aircraft in World War II.[32]

CHAPTER 3

FALL 1914 CAMPAIGN

☆ ☆ ☆ ☆ ☆

THE PREVIOUS CHAPTER ended on a sour note. It showed the First Corps Squadron ignored and mistreated and the Eleventh Corps Squadron having literally struck down an Austrian airplane in a battering-ram maneuver (at the dreadful cost of killing one of the Russian empire's best military pilots). By contrast, Petr Nesterov's friend, Viacheslav M. Tkachev, enjoyed a long life and a great career in military aviation during the Great War. Tkachev rose from squadron leader to division leader to inspector of army aviation on the Southwestern Front. After the first Revolution in 1917, Major General Tkachev held the title of chief field administrator for aviation and aeronautics, replacing Grand Duke Aleksandr, who carried the politically incorrect family name of Romanov. The son of a Cossack, Tkachev had been a junior officer in a Kuban Cossack Battery. His father's stature enabled Tkachev to take flying lessons with civilians in the Odessa Aero Club, followed by military flight training at the Sevastopol' Aviation School, where he earned the designation of Military Pilot in 1912. The next year he joined Nesterov in providing air reconnaissance for army training maneuvers held near Kiev. In October 1913 he made a lengthy training flight of close to one thousand miles, from Kiev to Ekaterinodar, with stops in Odessa, Kerch, and

Taman. At that point, he had achieved the Cossack rank of sub-captain (*pod'esaul*).[1]

The following March, Tkachev became commander of the Twentieth Corps Squadron, attached to the Fourth Army. With his squadron he participated in combat training maneuvers that spring. When war began in August the unit was well prepared for action. Unfortunately, the delay in transporting the squadron from its headquarters in Kiev to Lublin by train meant that when Russian troops entered the Austro-Hungarian Empire the squadron was still getting established at a field southwest of Lublin. At this crucial moment the cavalry failed to reconnoiter the army's forward passage, which was obscured by heavy woods and marshy turf. On August 23 (O.S.) two divisions that formed the substance of the Fourteenth Corps led the Fourth Army into enemy lands. The Russian divisions were stretched over a twenty-mile area, blind to the enemy's location, and were surprised to confront five divisions of the Austrian First Army piecemeal.[2]

It was bad luck for the spread-out Russians, who had stumbled into a concentration of a much larger Austrian force. Their troops took a beating and suffered serious losses. Two days later a preliminary reconnaissance flight by Tkachev on the Fourth Army's right flank revealed the approach of the Austrian First Army. Tkachev's altitude was low enough that several bullets from enemy troops on the ground hit his plane. One round damaged a fuel line. His sputtering aircraft soon turned into a glider, but he made it back and landed safely not far from Russian lines. The intelligence that Tkachev provided spared his corps and the Fourth Army from a devastating surprise attack on its flank. Subsequently, 4 Russian armies battled 4 Austrian armies. The Austrians endured defeat with 400,000 casualties, which represented nearly half their force; they also lost 100,000 prisoners. Clearly, Russian military leaders had made a terrible mistake by not devoting most of their military resources to the immediate defeat of Germany's ally. Regardless, the victory shoved Austro-Hungarian soldiers a good 100 miles back into their own territory and left Vienna pleading with Berlin for German help. The only good news for the Austrians was that the Russians had incurred major losses of their own that forced them to pause, recuperate, resupply, and rebuild their forces.[3]

Unlike the situation on the Southwestern Front against Austria-Hungary, Russian armies and air reconnaissance units did not enjoy such success in their invasion of East Prussia. Alarmingly, the Russians forces there lost despite fielding twice the forces. General Pavel Rennenkampf led the Russian First Army into German territory from the east and General Aleksandr Samsonov commanded the Russian Second Army, which invaded East Prussia from the south. General Maximilian von Prittwitz und Gaffron, in charge of the German Eighth Army, immediately recognized the Russians' intention to use their two armies as a pincer to squeeze the Eighth Army into oblivion. Prittwitz told the high command that he wanted to retreat west and across the Vistula River, turning East Prussia over to Samsonov and Rennenkampf. But the request infuriated General Helmuth von Moltke, the German chief of staff, who transferred two corps to the east and replaced Prittwitz with General Paul von Hindenburg, pulled from retirement, and General Erich F. W. Ludendorff, who earlier had audaciously taken the Belgian citadel at Liège. By the time the pair arrived on the scene, they discovered that a staff member, Colonel (soon to be general) Max von Hoffmann, had recognized that the Eighth Army could take advantage of the serious gap in the Russian armies. German pilots and observers confirmed the significant separation of Russian troops. Hindenburg would later comment that "without the airmen" there would have been no German victory.[4]

Besides aircraft, Hoffmann and the Eighth Army also had a second source of intelligence: the Russians were sending uncoded radio messages that revealed information about their supply problems and plans for movement. Their radio transmission errors lay somewhere between the absurd and the unbelievable. When Hindenburg and Ludendorff took command of the Eighth Army on August 23 (N.S.), Ludendorff developed a brilliant but smaller version of the Schlieffen Plan that actually worked. Aircraft and radio air waves confirmed that Rennenkampf had halted his advance because he needed supplies, and Stavka (the Russian military headquarters) ordered him to protect his right flank to the north by laying siege to the German citadel in Königsberg. It seemed a silly order, since the fortress was no immediate threat. Rennenkampf's first and only goal should have been to move quickly west and join Samsonov's army in destroying the major German force rather than worrying about the much smaller contingent of soldiers in Königsberg. Regardless, aviation and radio intelligence enabled most of the Eighth Army to move south and engage the Russian Second Army.[5]

As this process began, the Germans took a gamble to move their Seventeenth Corps, commanded by General August von Mackensen, southwest from its screening position in front of Rennenkampf's soldiers. Russian reconnaissance flights had noted the movement of the Seventeenth Corps, but Samsonov chose to ignore aviation scouting reports. He was one of those Russian generals who dismissed airplanes as playthings for civilians, unworthy of serious military work. Some Russian aviation historians claim that Samsonov and a major portion of his army could have escaped the German encirclement if the general had heeded the reports of Russian pilots and observers. Instead, the Battle of Tannenberg on August 29 (N.S.) led to the capture of almost 100,000 Russian soldiers, and Samsonov committed suicide. Bolstered by 2 transferred corps, the Eighth Army then moved east to engage and evict the Russian First Army from East Prussia.[6]

All too often, general officers in charge of corps-level aviation squadrons shared Samsonov's attitude toward aircraft. This, in turn, was one of several reasons that the reconnaissance task of Russian planes had such limited success in the fall of 1914. Another complication arose from the age of the Russian aircraft. Older planes were able to fly only short distances, and their combat range included the round trip plus time over the enemy's position. Because of this, in some instances at the start of the conflict on the Eastern Front the Russian cavalry had as good a chance as some aircraft in securing intelligence on German or Austrian troop movements. Furthermore, Russian planes carrying both a pilot and an observer—such as Deperdussin TT, Farman XX, and Voisin LA—were normally powered by 90-hp engines. Loaded with two crew members, such planes took more than half an hour to reach an altitude above 3,000 feet. That meant that if the plane gradually achieved 60 mph, and if the enemy troops were less than 25 miles from the airfield, the Russian plane might well have to fly low enough to suffer lethal ground fire from soldiers with rifles and machine guns. For this reason, pilots often flew two-cockpit planes without the observer so as to gain a safe altitude before approaching enemy lines. More often than not, scouting information gathered by the lone pilot proved to be much less accurate.[7]

Air reconnaissance performance also was weakened and disrupted by the unpredictable behavior and frequent loss of Russian aircraft. Belligerent countries generally matched Russia in operating delicate aircraft, although they had newer models. Planes had to be lightweight in 1914 because motors often had a modest rating of somewhere between 80 and 100 hp. This meant that the frame of the plane had to have the least amount of wood needed to keep the machine functioning in flight. That the airframes were made of wood explains why they were built by craftsmen rather than by machinists. Naturally there were specifications for the craftsmen to follow in the repetitious production process. Gradually the manufacturing scheme improved as companies began to apply templates to various aircraft components to improve and standardize parts in the assembly of aircraft under serial production. The Dukh Company, for example, used template boards to mark the outlines of longerons, cockpits, and fuselages. On the other hand, because experience and skill differed widely among workers building spars, ribs, longerons, and struts as well as frames for elevators, rudders, and ailerons, there could be slight and sometimes significant variations in serially manufactured aircraft. Combined with the low octane rating of gasoline, which bounced between forty and seventy, identical plane models often had different performance levels. Pilots had to fly a new plane a couple of times to measure the machine's speed, rate of climb, and response of control surfaces.[8]

Woe was the squadron that did not have at least one good engine mechanic. Most small planes were propelled by a rotary motor. Unlike modern reciprocating airplane engines that can fly thousands of hours before a major overhaul, the rotary engine had to be pulled from the airframe and disassembled for complete service after something like twenty-five hours of flight. Each squadron also needed an outstanding rigger who could repair and sew or patch the cloth exterior covering the wooden airframe as well as replace, rebuild, or repair control surfaces. Finally, support personnel with the squadron needed to find shelter for aircraft when nasty weather approached. At a minimum, the plane in an open field had to be securely tied down facing the wind, with control surfaces aligned with wing, tailplane, and rudder, or better yet locked or protected in some fashion. Failure to shelter or secure aircraft in a strong wind almost certainly would lead to airplane damage or even destruction. The fragile nature of aircraft in 1914 did nothing to improve their value among certain Russian generals. To be sure, aero engines would emerge more powerful and reliable in the near future, making possible the design and construction of heavier, more robust aircraft.[9]

Another problem that limited the effectiveness of reconnaissance aircraft was the fact the Russians had so many different types of planes. Their inventory included at least sixteen aircraft models, powered by twelve different engines that in turn were imported from five different countries. Neither the Aeronautics Unit (Section) before the war nor Grand Duke Aleksandr, as chief inspector of the Air Force, gave much thought to the problems posed by multiple machines. Aviation leaders should have focused on one or two top-performing small, single-engine planes that could easily be improved over time by designing advanced models that were more aerodynamic, better constructed, and powered by engines with higher horsepower. In addition, someone in authority should have insisted on both standardized parts and a bountiful supply of replacement parts. With frequent damage from combat and equipment failure stemming from age or poor construction of aircraft, the shortage of replacement parts caused a stunning rate of depletion. No one had planned to make replacement parts because the combatants had never before used airplanes in combat roles. Whatever the reason, by September Russia had lost 40 percent of its air assets. By the following month 16 squadrons from 4 armies had lost the use of 91 of their 99 planes. So critical was the situation on the Southwestern Front, for example, that General Aleksei Brusilov's Russian Eighth Army, which truly valued air reconnaissance, temporarily hired civilians and their aircraft at 300 rubles a month—equal to 150 U.S. dollars then, or almost 3,700 dollars a month in today's money—as replacements.[10]

The military also made the mistake of not educating peasant soldiers about the existence of the empire's air arm. Certainly, some troops in regular units had witnessed Russian aircraft in military peacetime maneuvers.

Many others, especially reserve units that had been mobilized in August 1914, had no idea that Russia possessed aircraft. Such infantrymen tended to conclude that only Germans could make contraptions that were able to fly. Occasionally, then, some soldiers regarded any man-made object in the sky as an enemy threat and took pleasure and pride in shooting at it. In the early phase of the war, Russian ground fire damaged Russian planes and even wounded fellow countrymen who were pilots. Naturally, these unfortunate actions only contributed to the limitations of Russian air reconnaissance in the fall. Time and experience gradually rectified such "friendly fire." In addition, many Russian aircraft were two years old, obsolete, and worn out from heavy use—especially the Nieuports that had been bought in 1912.[11]

No one had considered the possibility that wartime activity by heavily used older aircraft like the Nieuports might make losses more likely. There were no backup planes as War Minister Sukhomlinov had suggested in 1911. The error was compounded by the widespread expectation (held by all belligerents) that the war would be over in time for soldiers to celebrate Christmas at home. Thus, during the first month of combat, the five major Russian aircraft builders dawdled along, making approximately thirty-five airplanes and five aero engines a month. The loss of 40 percent of Russia's military aircraft in August showed clearly that airframe workshops needed to triple production if only to replace the planes that were being damaged or destroyed. The acquisition of engines produced by French subsidiaries in Russia or imported from abroad had to jump twenty-fold. The tsarist government soon placed orders for more domestically built aircraft, but it took time for Russian airframe manufacturers to train new workers, secure tools, acquire essential materials such as wood and fabric, and expand or move to new and larger production facilities. By October such delays had forced the tsarist government to place orders abroad for airplanes and engines.[12]

Imports of any kind presented their own set of problems, however. Before the war, most goods from the West either entered Russia by train from or through Germany or arrived by ship to Russian ports on the Baltic shore or in the gulfs of Riga and Finland, which extended from the sea. As one could expect, after August 1 train traffic ended and the German navy blockaded ship transports that had steamed to Russia via the Baltic, which became in essence a closed inland lake. By the end of October only one of Russia's gateway cities in Europe continued to be engaged in foreign commerce. The lone exception, Arkhangel'sk, remained closed to sea traffic during the winter and early spring months because of ice; it was located on the southern shore of the White Sea, which was mixed with fresh water from local rivers. Even in peacetime, seamen found the voyage challenging due to the bad weather and icebergs. To reach the Russian port, ships had to sail above the northern shore of the Scandinavian Peninsula and in the Arctic Ocean. During the war the trip became downright dangerous as German submarines intercepted and torpedoed some Allied ships that were attempting the northern run. U-boats sank the French-chartered *King David* and its cargo of Nieuport and Farman planes as well as the British transport *Variol*, which carried Sopwith aircraft.[13]

Besides placing orders for planes and motors with France and later with England, the tsarist government negotiated in October to purchase ten flying boats, fifty Model J biplanes (precursor of the famous Model JN or Jenny), and some eight-cylinder engines from the Curtiss Aeroplane and Motor Company. The U.S. firm shipped the items to Vladivostok, the only major Russian port that was open year-round. Located by the Sea of Japan, the port was also the terminus for the Trans-Siberian Railroad. Unlike the case with Arkhangel'sk, vessels sailing to Vladivostok did not have to worry about U-boats; the port was a safe haven for ships carrying U.S. and Canadian war materiel from west coast North America. Eventually, the United States lent 2.112 billion dollars to Britain and France, more than half of which went to Russia to buy North American products. But that meant that Vladivostok would be overwhelmed with goods because of the limited capacity of the largely single-tracked Trans-Siberian Railroad to transfer goods to European Russia—a setback that meant long delays for imported war products that had to travel more than six thousand miles to Petrograd before being distributed to appropriate military units.[14]

The focus on Vladivostok as tsarist Russia's most important, though flawed, port came because the Black Sea soon joined the Baltic in being blocked from international trade. The end of Black Sea commerce proved to be the most costly to the empire. A port city such as Odessa could have supplied France with much-needed wheat to feed its people and troops. In return, Russia might have received French aircraft, aero engines, and other military hardware without going so heavily into debt to its Allied partners. Ultimately, Germany closed the Dardanelles and Bosporus Straits between the Aegean and Black Seas, preventing all merchant vessels from carrying goods to and from Russia. On August 10 (N.S.), to avoid attack by the British Royal Navy, the German battle cruiser *Goeben* and light cruiser *Breslau* fled the Mediterranean and Aegean Seas and sought safety by passing through the Dardanelles and into the port of Constantinople, capital of the Ottoman Empire.[15]

The Germans had nurtured friendship with the Ottoman Empire for several years. Early on they agreed to finance the construction of a railroad across Anatolia to Baghdad. The friendship was enhanced when the Germans turned the two cruisers over to the Ottoman navy. The *Goeben* became the *Sultan Selim* and the *Breslau* was renamed *Midilli*. These newly acquired ships hoisted the Turkish flag and signed on some Turkish sailors for training. The *Goeben* became a Turkish warship over time and served the future Republic of Turkey past World War II. During World War I, however, the Ottoman navy's control over the former German ships was largely a sham. German crews remained in charge and both vessels came under the command of the Imperial German navy's Rear Admiral Wilhelm Souchon. Officially Souchon was a vice admiral in the Ottoman navy. Future reference to the two vessels will use their original names, since the ships still belonged to Kaiser Wilhelm II.[16]

On October 27 (N.S.) the *Goeben* and *Breslau*, accompanied by the Turkish cruiser *Hamidiye* along with several minelayers and torpedo gunboats, steamed through the Bosporus Strait and into the Black Sea. Earlier, Vice Admiral Andrei Eberhardt had wanted to take a task force into the Sea of Marmara and attack the two German warships near Constantinople. At a minimum, he wanted to keep some naval vessels near the Black Sea entrance to the Bosporus Strait so he could block the German warships from sailing into the Black Sea. Stavka, however, had decided that while the Ottoman Empire disliked Russia, it was neither anti-British nor anti-French. Since Stavka believed that the Turks would likely remain neutral, and since Russia did not need a third enemy, the General Staff ordered the Russian Black Sea Fleet to stay in or near its home ports. Stavka wanted to be sure that the Imperial Russian Navy did not provoke war with the Ottoman Empire. (Stavka was dominated by army leaders, whose wishful thinking led the command to make the wrong decision; the navy was weakly represented.)[17]

Before sunrise on October 29 (N.S.) Turkish torpedo boats launched their ordnance into the Odessa harbor, sinking 1 Russian gunboat and damaging several other vessels. The attack meant war. After dawn, when the *Goeben*, Turkish gunboats, and a minelayer approached Crimea, Junior Lieutenant Svetukhin, flying a Curtiss #10, sighted them and took note of the number of ships in the group, but lacking a radio, he could not report his observations until after he returned to base. (Russia imported some aircraft radio sets over time, but did not manufacture them domestically until late in 1916.) Several pilots later conducted some 20 reconnaissance flights during the battle. At this point the planes had neither bomb racks nor machine guns. Their pilots could only perform scouting duties to watch for the appearance of other enemy naval squadrons and mark the work of the Turkish minelayer *Nilufer*, which dropped a series of mines in the waters around the entrance to the Sevastopol' harbor. Once the *Goeben* was in range of the port, she fired shells at ships and facilities in the harbor, and Russian batteries were ready to respond. Sevastopol' had been fully warned about the Turko-German fleet from radio transmissions from 2 civilian vessels in the Black Sea as well as personnel in Odessa. Hence, the *Goeben* soon suffered damaging hits from Russian shells that, among other things, knocked out 1 of her funnels. The cruiser was forced to retreat before she could cause serious harm to Russian ships or buildings. By contrast, farther east, *Breslau* and a Turkish gunboat hammered the unprotected port of Novorossisk with 308 shells during a leisurely 2-hour assault.[18]

This attack on several Russian ports by ships carrying Turkish flags left the tsarist government with no choice but to declare war on the Ottoman Empire, which it did on November 2 (N.S.). Under wartime conditions, the action around Sevastopol' confirmed both the actual and potential usefulness of airplanes. Immediately, Vice Admiral Eberhardt ordered a daily program of air reconnaissance to help alert the Russian fleet to the presence of Turko-German ships. Additionally, bomb racks were installed on Curtiss planes to hold four relatively small bombs. At dawn on November 17 (N.S.), as Russian and Turkish troops began fighting in the Caucasus, a Russian naval task force bombarded the Turkish port of Trebizond. Located on the eastern shore of Anatolia, the city was the port of entry for Turkish soldiers and military supplies for the new Caucasian Front. The attack was communicated to Constantinople as the first Russian shells exploded on Trebizond.

Admiral Souchon decided that the repaired *Goeben*, joined by *Breslau*, should punish the Russian naval squadron as it returned to Sevastopol'.[19] The two cruisers steamed out of the Bosporus in mid-afternoon with hopes of intercepting the Russian ships somewhere below the Crimea around noon on November 18. Russian air reconnaissance planes from Sevastopol' spotted the enemy about twenty miles southeast of Ialta and returned to base. Quickly, seven Curtiss planes were loaded with light bombs and took off. By the time the planes approached the enemy cruisers Admiral Eberhardt's flagship, the *Saint Evstafi*, and other Russian ships were engaged with the *Goeben*. The Curtiss aircraft avoided the fire between these vessels by harassing the *Breslau*. Without bombsights and adequate practice dropping bombs during flight, it is likely that few bombs, if any, struck the ship. Nevertheless, there would be little serious damage from the small bombs. During the 14-minute battle, the *Saint Evstafi* took 4 hits, but the Russian ships delivered 14 shells to the *Goeben*, causing damage and killing 9 officers and 105 crewmen. Intelligence gathered later in Constantinople revealed the casualties. As faster ships, the enemy cruisers escaped southeast into a fog bank. Tsar Nikolai II later awarded Admiral Eberhardt the Order of the White Eagle, with Crossed Swords, for the naval victory.[20]

While the Black Sea Fleet successfully employed aircraft the moment war began, Igor Sikorsky's four-engine reconnaissance-bomber was left out of the action. The plane figuratively had its ups and downs. After the Riga-based Russo-Baltic Wagon Company hired Sikorsky in 1912, he worked on his model S-6 biplane with craftsmen, pilots, and engine mechanics in the firm's Petrograd branch. Later he headed the Petrograd subsidiary and designed many other small aircraft. His S-10, for example, had floats (pontoons) and ended up with the Baltic Fleet. Except for the S-16, most of his small planes were manufactured in limited numbers. Thanks in part to

Grand Duke Aleksandr, military aviators favored foreign planes—especially French designs—even if they were Russian copies. But in 1912 Sikorsky had a lengthy talk with company chairman Mikhail V. Shidlovskii about the designer's hopes of building a large, multi-engine plane.[21]

Sikorsky argued that a bigger plane could fly farther and carry passengers and goods or military ordnance. But the impetus for his vision of larger aircraft emerged in 1911, when he flew his first really successful airplane, the S-5. During an exhibition flight, a mosquito had accidentally entered the fuel system and became lodged in the tiny jet of the carburetor. He had just taken off and was barely above some tall trees when his engine quit. Unfortunately, he was flying too low to avoid the trees if he turned around and tried to return to the field that he had just left, and he had to make an emergency landing in a railroad yard that was partially enclosed by a stone wall. When he landed, his plane flipped over and stopped shortly before the wall. Because he came within feet of a fatal end to his career in aviation, Sikorsky strongly believed multiple engines made flying safer. It should come as no surprise, then, that he designed a multi-engine aircraft, which would enable an engine mechanic to leave the cabin during a flight and go out on the wing to repair a malfunctioning motor.[22]

Fortunately for Sikorsky, Shidlovskii was a true entrepreneur. Born a member of the landed aristocracy, he had served as an officer in the navy and later in a high administrative position in the empire's Ministry of Finance. After the 1905 Revolution Tsar Nikolai II managed to retain his position as emperor, in part by allowing the formation of a national elected assembly, or Duma, with limited powers. Eventually, Shidlovskii joined the State Council, which was dominated by the wealthy and nobility. The council was legally authorized to veto Duma legislation. Despite Shidlovskii's background and official ties to the tsarist government, he was not such a traditionalist as to ignore new and progressive business opportunities in his role as head of the Russo-Baltic Wagon Company. His firm built railroad cars for export, not just domestic consumption; it also became the first in the empire to manufacture an automobile, to establish a major aviation factory in Saint Petersburg, and to build the first Russian-made aero engine. No surprise, Shidlovskii enthusiastically agreed to fund Sikorsky and supply him with craftsmen, mechanics, and pilots to help him build and test the world's largest aircraft. Sikorsky soon took charge of the company's Saint Petersburg subsidiary, where his large and small planes would be built.[23]

Sikorsky had worked on the biplane's design for a year before August 30, 1912 (N.S.), when construction began. Although the aircraft carried a couple of different names during the construction process, the team that built the airframe entitled it *Grand*—French for large. Remarkably, without the benefit of later knowledge of aeronautical science, Sikorsky used intuition to create the eighty-eight-foot-long, narrow wings with a high aspect ratio that ultimately enabled the huge, underpowered craft to fly. Many small planes of the time had a fuselage that was kept in the air by short, wide, low-aspect-ratio wings. The major sections of the *Grand* were finished in April 1913, and the frame was transported north to the military's Komendantskii Aerodrome. There the aircraft was assembled, supplied with dual controls and homemade speed and bank indicators, and equipped with four water-cooled Argus motors rated at 100 hp each. A balcony served as the nose of the plane. Farther back in the sixty-five-foot fuselage a door opened into the pilot compartment, followed by another door to the passenger cabin, which was furnished with four chairs, a sofa, and a table. The passenger section also had a water closet, clothes closet, and two side doors that permitted an engine mechanic to walk out on the lower wing to service a motor during flight.[24]

The plane was finished and furnished by the second week of May. At 10 p.m. on May 13 (O.S.), when the northern "White Nights" still provided some light, Sikorsky piloted the *Grand* for a short ten-minute test flight with an engine mechanic on board. It lasted just long enough for the pilot to reach an altitude above six hundred feet, do a couple of turns, and perform a practice landing in the air before returning to the field. The test flight proved wonderfully successful. Everyone at the Komendantskii Aerodrome, including neighbors from the surrounding area, knew about the *Grand*'s upcoming first flight. Hundreds or—as Sikorsky wrote years later in his autobiography—thousands of jubilant spectators ran onto the field and carried him to a nearby hangar, where a smiling Shidlovskii received Sikorsky's glowing report on the aircraft's performance. After that, Sikorsky conducted fifty-one more flights, often with seven or eight passengers, that made the huge airplane a common sight in the skies over Saint Petersburg. At the request of Tsar Nikolai II one of those flights took him twenty-five miles to an army field close to Krasnoe Selo. The emperor was so impressed by the plane and its pilot-designer that a couple of days later he sent Sikorsky a gold pocket watch with an image of the imperial Russian eagle mounted on the back.[25]

In late summer 1913 the army conducted aircraft competitions, and two Sikorsky small planes were entered in the contests. The *Grand* was stored outside the hangar that protected them, and an 8-foot fence surrounded and guarded the large machine from curious visitors at the Komendantskii Aerodrome. Unfortunately, one of the competing planes, an airborne Voisin pusher type that was on approach toward the area where the *Grand* was parked, encountered a problem. Wooden supports in the small airborne plane cracked and splintered around the mounted engine. The motor then simply fell out of the craft, smashing the double wings on one side of the *Grand*. Amazingly, the pilot who lost his engine survived and was even able to walk away from his plane, which had crash-landed. Sikorsky, who witnessed the event, chose not to repair the damaged *Grand*, but rather scavenge parts from it. Sikorski reasoned that doing so would give him and his team a head start in building a slightly bigger four-engine plane, the Il'ia Muromets. The larger aircraft was finished in mid-winter and equipped with skis instead of wheels for landing gear. A second Il'ia Muromets was completed in the spring of 1914, with more powerful engines that together generated 530 hp, and was capable of flying to an altitude approaching 7,000 feet.[26]

In the second Il'ia Muromets, Sikorsky conducted additional local flights carrying government officials, military personnel, and Duma representatives. Thus he felt comfortable to plan a long-distance flight from Saint Petersburg. On June 30, 1914 (N.S.), accompanied by pilots Lieutenant Georgii I. Lavrov and Staff-Captain Khristianskii F. Prussis and engine mechanic Vladimir D. Panasiuk, Sikorsky took the second Il'ia Muromets on an amazing journey from the empire's capital city to his hometown of Kiev. The flight had its moments of joys and challenges. And, yes, the mechanic did have to go out on the wing during flight—in this case, to put out a fire that had been caused by a ruptured fuel line. That problem forced an unscheduled emergency landing, but the approximately 750-mile flight from Kiev back to Petrograd required only one planned stop for gasoline. Clearly even a loaded Il'ia Muromets would have a safe flight range of more than 325 miles. Tsar Nikolai II awarded Sikorsky the Order of Saint Vladimir, Fourth Class, for service to the state with honor. The aircraft designer also received an award of 100,000 rubles (about 50,000 dollars) to continue his work on refining the Il'ia Muromets; Shidlovskii's company also gained a War Ministry contract to build ten more copies of the machine.[27]

The military contract for 10 aircraft did not come solely because Sikorsky had piloted a 1,500-mile trip in a big airplane. Threats of war overshadowed and tended to hide what should have been worldwide fame for Sikorsky and his aircraft. His flight set international aviation records for distance, duration, and altitude. After a Serbian nationalist murdered the heir to the Austro-Hungarian throne, that empire secured a blank check from its German ally. Kaiser Wilhelm II would accept whatever Austria-Hungary needed to discipline Serbia, which it blamed for the assassination. Vienna handed Belgrade a severe ultimatum that in compliance would have violated Serbia's sovereignty. Austria wanted war and found Serbia's conciliatory response unacceptable. Because of the system of military alliances, when the empire declared war on the small Slavic country what should have been the Third Balkan War horrifically and quickly expanded into the Great War.[28]

As these events unfolded, Sikorsky, assisted by Anatolii A. Serbrennikov, busily redesigned and, with the help of some three hundred workers, began constructing a new version of the Il'ia Muromets. Sikorsky was not clueless about the changes that needed to be made. His background as a student in the Imperial Russian Naval Academy and the several years that he spent with officers at military airfields helped him understand that the third and future versions of the large plane had to be substantially different from the first one. Gone were the chairs, table, sofa, and other comforting features. Sikorsky then added space for military equipment and bomb racks. He also narrowed, lightened, and streamlined the fuselage in order to reduce drag, improve speed, and expand the flight ceiling above the reach of small-arms fire from the ground. He made plans to install some type of metal armor plates near the front end of the cabin to keep the pilot safe during combat missions. The fuel tank above each engine received special packaging to prevent the loss of fuel from an enemy bullet. It was a forerunner of the self-sealing tanks that were so widely used during World War II. For additional protection, some later models of the Il'ia Muromets held four fuel tanks, one for each of the four engines on top of the fuselage.[29]

It took only seven weeks for the third Il'ia Muromets to be produced as the Model V. (In the Cyrillic alphabet, "V"—pronounced "veh"—is the third letter in the alphabet and printed in Cyrillic as "B.") The company built several copies of the third version. In wartime circumstances, the Petrograd factory stayed open twenty-four hours a day. Around-the-clock shifts of craftsmen built the plane's skeletal structure with woods of spruce and pine, fashioned into longerons, struts, ribs, spars, formers, stringers, and plywood panels and then braced with internal and external wires. Fabric covered the airframe. Sikorsky worked long hours and into the night. His intuitive powers remained strong. He felt, for example, that the loads and stresses on the tail structure, especially the horizontal stabilizer or tail-plane, required more strength. Aeronautical science later confirmed his intuition. During the day he spent time with and assigned tasks to company engineers. He also inspected and supervised the work of those constructing and finishing the aircraft.[30]

Sikorsky had one other important task related to the Il'ia Muromets. The army quite naturally expected pilots for the ten airplanes on order with the Russo-Baltic Wagon Company. Sikorsky, then, had to set aside some time during the day to supervise a pilot-training program. To aid Sikorsky in this process, the military assigned to him one of its best and most successful flight-training instructors, Staff-Captain Evgenii V. Rudnev. As the author of aircraft guides for student pilots, Rudnev quickly and easily became competent to fly the Il'ia Muromets. Hence he took on the responsibility of training other experienced pilots who transitioned from flying small, single-engine planes to the giant four-engine aircraft. There was one problem with Rudnev that was not immediately evident to Sikorsky, however. Rudnev had spent four years flying small aircraft. He loved their relatively quick response in making turns and gaining or losing altitude. Moreover, like most army pilots, he favored French-designed aircraft and held suspect all planes conceived by Russians.[31]

In mid-September 1914 the army decided it was time to see what the new version of the Il'ia Muromets could do in a combat reconnaissance mission. It ordered Rudnev to take one of the two original Il'ia Muromets aircraft from Korpusnoi Aerodrome near Petrograd and fly it to Belostok, which lay approximately 140 miles northeast of Warsaw, not far from the battle zone between Russian and German forces. The September flight was around 150 miles shorter than the one in July, which traveled from Kiev to Petrograd. Still, even with a stop for gasoline at Novo-Sokolniki, the flight north from Kiev took about 13 hours. By contrast, the shorter September plane flight from Petrograd arrived in Belostok in October, 23 days after its departure. When the Il'ia Muromets took off from the Korpusnoi Aerodrome, it encountered headwinds that reduced the huge craft's speed at times to a low of 17 mph. The aircraft ran out of fuel and forced-landed, causing damage to the landing gear, which had to be replaced. Once the plane was airborne again, one engine stopped running, requiring another emergency landing—this time to repair and overhaul an engine. After the plane finally arrived at Belostok, the power plant problems continued, leading to delays and making it difficult for the plane to reach an altitude that would be safely above ground-fire from enemy troops. Rudnev considered the Il'ia Muromets unreliable and asked the headquarters staff of the Northwestern Front for two small planes to conduct the air reconnaissance originally assigned to the Il'ia Muromets. Unfortunately for Sikorsky and the Russo-Baltic Wagon Company, the situation actually got worse.[32]

CHAPTER 4

NEW ROLES FOR AIRCRAFT IN 1915

☆ ☆ ☆ ☆

IN THE FALL OF 1914, as Sikorsky was building the first two advanced military versions of the Il'ia Muromets, an earlier-generation model of the large plane took off from Petrograd to carry out a demonstration reconnaissance mission at the Northwestern Front. Piloted by Staff-Captain Aleksei V. Pankrat'ev, the four-engine behemoth had to make a forced emergency landing after having been shot at by Russian soldiers. Pankrat'ev and his crew decided to complete their journey to Brest-Litovsk by dismantling major portions of the aircraft and transporting them by train to Brest-Litovsk. Considering what had happened to IM-1 and IM-2, it was understandable that the headquarters staff of the Northwestern Front chose not to accept any Il'ia Muromets aircraft. The large plane seemed to be totally unsuitable for use in an active combat zone. Grand Duke Aleksandr, the field inspector general of aviation, strongly endorsed the headquarters staff conclusion, which Stavka then accepted as well. Apparently Sikorsky's amazing aircraft would not be used by the military.[1]

It was not exactly surprising that Sikorsky and his employer decided to challenge the army's rejection of the large plane. The new streamlined Model V enjoyed a better performance level that enabled the aircraft to cruise over enemy troops at higher and safer altitudes. With two improved airplanes in hand, Sikorsky and Shidlovskii collaborated on a document that Shidlovskii was to sign and submit to War Minister Sukhomlinov, who in turn would share it with Tsar Nikolai II. This was not some type of obsequious, pleading request. It proved to be a stunning recommendation for a program that took orders not from some corps, army, or front, but only from Stavka. Shidlovskii could send the recommendation to the highest levels of government with some assurance that it would be read with care. He had served the tsarist regime in prominent positions and headed one of Russia's largest and most advanced manufacturing companies. The missive called for the formation of a special aircraft detachment named the Eskadra Vozdushnikh Korablei (Squadron of Flying Ships), identified commonly by its initials, EVK.[2]

Moreover, with Sikorsky's help, Shidlovskii specified that the proposed EVK should have its own workshops, training school, photographic section, storage facilities, and flying missions on behalf of Stavka. And he volunteered to take the leadership of the detachment. The favorable response of the tsar and war minister resulted in the commissioning of the former naval officer as a major general in the Imperial Russian Army. According to Sikorsky, Tsar Nikolai II ordered the promotion (despite claims by Konstantin N. Finne, the squadron's surgeon, who wrote the history of the EVK, that the increase in rank actually emanated from Stavka). For his part, Grand Duke Aleksandr opposed both Shidlovskii's new promotion and even the EVK's existence. The tsar's brother-in-law remained unshakably faithful to the value of small planes of French design. Shidlovskii and his employee, Sikorsky, arrived at the EVK's headquarters at Iablonna in January 1915. The new base was situated near the Vistula River, about halfway between Warsaw and the Novogeorgievsk Fortress to the north-northwest. Fortuitously, the two revised Il'ia Muromets aircraft and two earlier versions were based less than forty-five miles east of the German Ninth Army and south of the German Twelfth Army.[3]

Sikorski oversaw the first test flights of the squadron's planes in January. Over time he and the detachment's mechanics repaired and fine-tuned the engines to ensure that the large planes would be ready for a series of training flights. Until mid-1916 Sikorsky had spent much of his time with the EVK headquarters, which moved around quite a bit that year. He personally flight-tested every new plane and worked with mechanics after each combat mission to repair or rectify any unsatisfactory aspect of each aircraft. Sikorski's dedication to the squadron may explain why his marriage to Olga Fyodorovna Simkovich failed, although their relationship produced a daughter, Tania, who grew up to become a college professor in the United States. The first flight without Sikorsky at the controls took place on January 26 (N.S.), with Captain Georgii Georgievich Gorshkov piloting the first Model V plane. Like most EVK pilots—and unlike many small-plane pilots—he lived through the Great War at least to the Russian Revolution. After April 1917, under Russia's new Provisional Government, War Minister Aleksandr I. Guchkov essentially fired General Shidlovskii, and newly promoted Colonel Gorshkov was appointed to command the EVK. Shidlovskii lived until 1919, when he was captured and executed by the Bolsheviks (Communists) for having fought on the wrong side during the Russian Civil War.[4]

On January 26, 1915 (N.S.), Gorshkov piloted the large plane south to the area around Warsaw. Twelve passengers, all future crew members, accompanied him on the flight as part of their training experience. They also served as ballast, to prove that the loaded plane—which usually contained bombs—could easily reach an altitude above 8,000 feet and thus be safe from the small-arms fire of enemy troops. The son of an Ural Cossack officer, Gorshkov graduated from a military cadet school, an engineering college, and the aeronaut training park at Volkhov Field. Assigned to a balloon detachment, he also served briefly as a dirigible commander. Gorshkov's extensive experience with flight led to his being tapped to go to France where, in the summer of 1911, he learned to fly a Blériot Model XI airplane. His several years as a pilot helped explain his selection in 1914 to become one of the first military pilots to familiarize himself with the cockpit of the Il'ia Muromets. For this reason he briefly commanded the EVK until Shidlovskii arrived.[5]

Weather permitting, the EVK spent a month learning, practicing, and flying a series of rehearsal combat missions to make certain that all crew members understood their roles. The practice sessions provided additional experience and promoted confidence in pilots; engine mechanics; artillery officers, who took photographs; and soldiers, who handled bombs and machine guns. On February 28 (N.S.) Gorshkov and his crew flew west near Plotsk, dropping six hundred pounds of bombs on German trenches and taking photographs of the enemy's positions. Despite Grand Duke Aleksandr's negative views, the successful mission quickly turned Stavka into a supporter of the EVK. After several combat missions, Gorshkov received the Order of Saint Vladimir, Fourth Class, with Swords. The order, initiated by Catherine the Great in 1782, was awarded "with swords" for military valor and achievement. In 1915, in what must have been galling to the grand duke, Stavka promoted Gorshkov to lieutenant colonel (*podpolkovnik*) at a time when pilots of small planes could never hold such a high rank. Ace pilot Aleksandr A. Kozakov was still staff-captain (*shtabs potmistr* for an officer who came from the cavalry) in 1917 when the grand duke was fired as leader of Russia's military aviation.[6]

Air reconnaissance allowed the Germans to know where the Russian large planes were located. Soon after Gorshkov and his crew safely returned to their Iablonna aerodrome, a single German plane flew over and dropped a small bomb on the airfield. That response would be repeated in the future, but with several planes and bigger bombs. The air attacks generally caused little or no damage to planes, facilities, or personnel. Because the early wartime Il'ia Muromets carried two Maxim machine guns and two Madsens, there was less concern about being strafed by German aircraft during flight. The only way an enemy plane could fire upon an Il'ia Muromets would be for the observer in the second cockpit to wield a light machine gun, preferably secured. (In later models the machine gun would be mounted on a Schneider Ring, which would enable the observer to train the weapon in different directions.) Regardless, the observer needed open space to fire the machine gun at Russian aircraft; otherwise, he might injure or kill his own pilot and damage his own plane. The net result meant that for a good shot on the unprotected side of the cockpit the German plane needed to fly parallel with the reconnaissance-bomber. That approach, however, would place the German plane in harm's way, since it could be riddled unmercifully with Russian bullets from at least two machine guns.[7]

Genuine fighter aircraft emerged in 1915. Just as they did in the case of balloons, dirigibles, and aircraft, the French liked to take responsibility for fighter aircraft. In 1913 Frenchman Roland Garros, who was the first pilot to fly across the Mediterranean Sea, added a metal deflector plate to the propeller of his Morane-Saulnier monoplane and had a Hotchkiss machine gun installed in front of his cockpit and behind the engine. The gun fired through the arc of the propeller. About 10 percent of the bullets hit the propeller, but bounced off the deflector plate. Until Garros' breakthrough, French aircraft had flown parallel to German planes and exchanged pistol or carbine bullets. Garros could now aim his Morane-Saulnier aircraft—and direct the machine gun perpendicular, up or down—thus enabling him to shoot at the enemy plane from any side. During the first weeks of April, Garros totally surprised three German pilots by firing his weapon through the arc of his propeller and shooting them down. Unfortunately for Garros, ground fire in turn brought his plane down on German-held territory on April 18 (N.S.).[8]

Garros ended up in a prison camp; his plane ended up with an impressed German military. The Germans asked Dutch aviation pioneer Anthony H. G. Fokker, who built engines and planes for the Germans at his factory at Schwerin east of Hamburg, to replicate the Garros plan for firing a machine gun through a rapidly rotating propeller. Fokker was appalled by the notion of using a deflector plate. A bullet could conceivably bounce back and hit the pilot or damage the engine. Repeated bullets striking the propeller might have an unwanted effect on it or the crankshaft. Logically, Fokker asked his team, led by Reinhold Platz, to synchronize machine gun fire somehow with the movement of the propeller. Several countries had experts already studying this problem. The solution that Fokker's team employed was a synchronized "interruption gear" that resembled the patented work of Franz Schneider used by the German aviation firm Luft-Verkehrs Gesellschaft (LVG). In May, Fokker applied the synchronized gear to the light but sturdy Eindecker monoplane that was built by his company.[9]

While not ignoring Garros' inspiration, most aviation historians acknowledge Fokker's Eindecker as the world's first fighter aircraft. On the other hand, Russia did form one of the earliest fighter detachments. At the beginning of the year, with German armies established north and west of Warsaw, the number of German reconnaissance flights over the city and surrounding areas began to increase noticeably. Lieutenant Nikolai S.

Voevodskii, an admiral's son whose father briefly headed the Imperial Naval Ministry, led a special squadron that focused on keeping enemy scouting aircraft away from the Warsaw area. He made no effort to alter Russian planes to create something akin to the German Eindecker. But he did have access to pusher aircraft, such as Voisin and Farman, that enabled the observer to man a swivel-mounted machine gun in the front of the aircraft. The detachment also had Nieuport 10 tractor biplanes, which could be equipped with a static machine gun on the top wing that fired bullets straight ahead but above the propeller's arc. As a result, Russia had assembled a squadron that specialized in shooting down enemy planes.[10]

Voevodskii, who died in Spain at the ripe old age of eighty-six, was an ace pilot who earned numerous awards, but for two reasons his success as a Russian military aviator did not lead to advances in fighter aircraft. First, most of the empire's light machine guns—Hotchkiss, Lewis, Madsen, and Maxim—had to be imported, and there were never enough of them available. Only because of Stavka's influence did EVK get top-priority access to all the machine guns it needed. The other problem was how to synchronize the machine guns to fire through propellers. It took time for Russia's western Allies to match and mimic Germany's accomplishment. In Russia, only the Moscow-based Dukh Company had mastered the synchronized machine gun—at least on the Le Rhône engine. Shortly after the February–March Revolution, it signed a navy contract to build seventy-five Nieuport pursuits, most of them Models 17 and 21, which also were used by Allied air forces such as the U.S. Army Air Service.[11]

It is not clear whether Dukh actually fulfilled the contract. As was the case in other shops, the first phase of the Russian Revolution caused workers to demand the eight-hour day and significant wage increases, followed by work stoppages to secure these demands, and, finally, factory takeovers by labor. Except for some imports manufactured by the French companies Société Pour L'Aviation et ses Dérivés (SPAD) and Morane-Saulnier in 1917, Russian fighter planes continued to be pusher or tractor biplanes with a machine gun mounted in the front cockpit or on the top wing. A variant of these was an upward-angled weapon positioned through the top wing. The trigger was right in front of the seated pilot, but the pilot was only able to pull that trigger when his plane was below the enemy's aircraft.

In a firefight with a German plane, the upward-angled machine gun was an extremely limiting feature. Under those circumstances the Russian pilot's best chance to survive and enjoy a victory came when he flew unseen from behind and below the enemy aircraft.[12]

Unlike the Eastern Front and Black Sea theaters of battle, for months the Baltic had little need for combat aircraft with guns and bombs. Nevertheless, Admiral Nikolai von Essen, who headed the Baltic Fleet, strongly supported the use of airplanes for the crucial reconnaissance role. Among other initiatives he endorsed the February 1915 conversion of the passenger vessel *Orlitsa* (female eagle) into a hydrocruiser (*gidrokreisera*) that carried up to nine aircraft on its top deck. Like similar ships in the Black Sea, the vessel was equipped with a crane so it could drop aircraft into the water and retrieve them after they had flown their missions. Later that spring the *Orlitsa* spent time around the islands (Dägo, Moon, Ösel, and Worm) at the entrance to the Gulf of Riga. The hydrocruiser's planes conducted scouting missions to provide naval vessels with intelligence on the approach or absence of German warships. Alone, the *Orlitsa* could carry more flying boats than the Baltic Fleet originally possessed. Essen eagerly requested—and finally received—several more FBAs and two early Grigorovich (M-2 and M-4) flying boats.[13]

The Russian mine offensive continued into the late fall of 1914. Among various German transports and warships, the cruisers *Augsburg, Friederich Carl,* and *Gazelle* fell victim to the mines, but only *Gazelle* had to be scrapped. Moreover, no German planes had the range to scout or attack Russian ships and aircraft. The closest German town and port to the tsarist empire, Memel, had been mined by the Russian navy and became unusable to the Germans. In fact, the Russian army briefly occupied the port. The Baltic froze over by the start of February and fresh-water rivers kept salinity levels in the region low. In 1915 the gulfs of Finland and Riga, which served as the base for Russian ships, remained iced over until early May; the ice and floating ice reduced German naval activity. In late January 1915, however, a harbinger of future activity occurred at Libau, a port on Russia's Courland coast: Russian forces shot down the German airship PL-19 as it approached Libau, and several German cruisers assigned to shell the port encountered mines or were grounded temporarily in shallow water.

As a result, Libau avoided serious damage and potentially a German invasion.[14]

On March 28, 1915, the first Russian task force in the Black Sea was formed around two hydrocruisers, *Almaz* (diamond) and *Nikolai*. The sea squadron included five dreadnoughts, two cruisers, ten destroyers, and six minesweepers. It steamed from Sevastopol' to the mouth of the Bosporus Strait, where the large ships shelled coastal Turkish targets such as lighthouses while seaplanes and flying boats bombed batteries and the torpedo boat *Samsun*. The task force moved east to the coal-producing regions of Anatolia. At this point, there was no rail line from the coal mines to the Ottoman Empire's capital city of Constantinople, and good roads and heavy-duty trucks belonged to the future. The only way to transfer the heavy, bulky mineral was by ship. Thus, while the Russian warships destroyed eleven coal-carrying Turkish vessels hydrocruiser aircraft bombed the port of Zonguldak, which was the point of entry for Turkish coal.[15]

Lieutenant Utgof received the prestigious Order of Saint George, Fourth Class, for his bravery and success during the mission. His bombing of a Turkish battery killed several enemy soldiers; his raid over Zonguldak caused a fire that scorched the city's electric generating plant and put it out of service. A graduate of the Saint Petersburg Polytechnic Institute and the Sevastopol' Aviation School, Utgof was later assigned to be an assistant military attaché at the Russian Embassy in Washington. The son of Russian nobility, he could not go home at war's end because of the class hatreds held by the new Communist government. He did well in America and started a farm near Roosevelt Field on Long Island. In 1923 his farm and its sheds became the home and workshop where Sikorsky lived while he designed and built his first airplane in the United States, the S-29-A twin-engine biplane. It was at Utgof's home that Sikorsky's sisters and his daughter, Tania, joined him after their perilous escape from Soviet Russia. Sikorsky's brother, Sergei, a naval officer, died during the fall of 1914 when a German torpedo from the submarine U-26 blew up the cruiser *Pallada* in which he had served in the Baltic Sea.[16]

Utgof's heroics and the success of the hydrocruiser task force in bombarding portions of the Ottoman Empire in the opening months of 1915 illustrate how aggressively the Imperial Russian Navy responded to the attack by the Central Powers. In fact, the navy surrendered nothing in either the Black or Baltic Seas, and it struck back forcefully with mines, shells, and bombs against any enemy assault or threat. By contrast, the scorecard for the Imperial Russian Army was mixed at best. On the Northwestern Front the army suffered a terrible defeat in the Battle of Tannenberg. The Russian Second Army lost in part because General Samsonov ignored the reports of air reconnaissance crews. A German offensive then pushed Russian forces out of East Prussia and attacked the tsar's troops on the left bank of the Vistula River. Even so, these German victories were modest and of little or no consequence strategically.[17]

The Russian army's scorecard for the Southwestern Front reflected a far better performance, partly because the enemy suffered from some major disadvantages. First, members of the Austro-Hungarian forces spoke fifteen different languages, burdening it with a polyglot of monumental proportions; a few units actually adopted English as the best common denominator. As the war dragged on many soldiers came to appreciate the possibility—even the hope—that the empire would collapse at the war's end and splinter into a group of individual countries. Even so, Russia's army more than matched the Austro-Hungarian force, which retreated westward in Galicia. Surrounded by the Russian Eleventh Army just before Easter, the Austrians surrendered the great Przemysl Fortress, which had run out of food and ammunition. The Russians took 120,000 prisoners of war in that confrontation and confiscated more than 1,000 pieces of artillery, including 8 12-inch mortars. At first, the great victory appeared to be far more significant than Germany's triumph at Tannenberg. The Russian Third and Eighth armies already in or near the passes of the Carpathian Mountains potentially could be augmented by the Russian Eleventh Army. This enlarged force would then be able to move across the Carpathians to the Hungarian plain. Then both capitals of the Austro-Hungarian Empire, Vienna and Budapest, would be open to Russian attack.[18]

Two factors highlighted the importance of Russian air power in 1915 and also saved Austria from a possible defeat. First, Brusilov's Eighth Army soldiers in the Carpathians lacked sufficient ammunition. Replacement troops arrived without rifles and bullets because supply depots were empty. The war materiel shortage—from bullets and shells to rifles and artillery—

had become desperate. Russia had not planned for a war that would last longer than a few months; indeed, during the first half of 1914 military orders had been cut so that the empire only produced forty-one new rifles by July of that year. Second, Russia did not provide for enough coal supplies to enable its factories to meet wartime needs. Many of Russia's military-related industries were located in the capital city. Far from domestic coal mines, several hundred factories and workshops relied on coal imported from England to produce power and heat for metalwork. When the war began coal imports almost ended. To add insult to injury, Russian coal miners were drafted into the army, so even internal fuel supplies fell off. War Minister Sukhomlinov drew sharp criticism for such widespread unpreparedness and eventually was dismissed and convicted of malfeasance in office. The Bolshevik Revolution enabled his escape from Russia.[19]

Reconnaissance aircraft played a critical part in the Russian army's survival. The shortage of shells made it more important to send aircraft to identify and pinpoint enemy targets. The result was to make each of the available artillery shells more effective in damaging the enemy's position. Normally, the scouting pilot or observer would drop a target message near the Russian artillery battery. Occasionally, a unit would work out visual signals—such as smoke bombs and streamers—for use between pilots and the ground crew; wireless communications did not become standard until a later war. A second development that heightened the importance of airpower was a decision by the German High Command to save its country's ally, the Austro-Hungarian Empire, by defeating Russia. The Western Front had turned into a stalemate of trench warfare, giving Germany the opportunity to transfer eight seasoned and well-equipped divisions to the Eastern Front into a new German Eleventh Army under General August von Mackensen. The new force assembled in April 1915 alongside the Austrian Fourth Army in western Galicia. Incredibly, the two armies were packed in the twenty-two-mile space between the city of Tarnow and the town of Gorlice. The Russian Third Army occupied the space opposite the two large forces of the Central Powers.[20]

But Petrograd's plans for a Russian offensive against Austria farther south drained all the reserve troops from the Third Army, leaving Russian generals with a bitter irony. Because the Russian army had been in a stationary area with few firefights, it had a decent supply of ammunition and weapons on hand. But its supply depots were empty, so the Third Army's quartermaster section could not replenish artillery, bullets, machine guns, rifles, and shells that might be used up, damaged, or lost in a major battle. Foreign markets belatedly helped Russia's weak industry. Russia received approximately half of some 2.112 billion dollars in U.S. loans to Great Britain and France. The tsarist government used the money to purchase war materiel in Canada, France, Great Britain, and the United States. In America, the Russians sought dozens of items, ranging from airplanes and automobiles to steam locomotives and surgical instruments. But not all got to Russia in time to aid the immediate conflict. For example, the United States built and sent 2.5 million rifles to the tsarist empire, but these did not arrive until 1916. The French received Russian orders for 586 aircraft and 1,730 aero engines, but only 250 airplanes and 268 motors actually arrived before the end of 1915.[21]

There were gaffes on the battlefield as well. In April 1915 General Radko Dmitriev, who commanded the Third Army, failed to order his corps and divisional generals to move artillery away from enemy view and to require soldiers to strengthen and deepen trenches. The army also needed fallback positions—deep trenches a thousand feet or more to the rear and linked by zigzag channels to the forward line. It may have been unfortunate that the area occupied by the Third Army over the winter had been remarkably quiet. Quite possibly the months of relative peace had lulled commanding officers into forgetting that they were engaged in a fearsome and deadly war. Unforgivably, the generals chose to ignore their own air reconnaissance reports that early in April had spotted clearly and correctly the start of a major German force buildup just across from Third Army lines. Indeed, a combination of pilots and observers, local civilians who had no love for the Austrians, and even Austrian deserters pinpointed the location and eventually the date of the enemy attack.[22]

On May 2, 1915 (N.S.), the German Eleventh Army started a massive and effective artillery shelling of the Russian Third Army. The bombardment eliminated the barbed wire in front of the Russians' shallow trench, which soon became the grave for thousands of soldiers. A number of Russian soldiers and support troops fled, but had no second or third tier of trenches where they could regroup and establish a defensive line against the enemy. Many would be shot or captured in the open country to the east. German and Austrian troops poured through the large undefended gap that had been carved out by artillery and left vacant by escaping Russians. In an effort to plug this growing hole, the Russians sent in two regiments and two cavalry divisions, but such counterattacks proved useless as these forces disappeared or melted away in the maw of war. Soon the breakthrough widened to twenty-eight miles as the Russians ran out of both options and ordnance to close off the gap. From the Romanian border in the south to the Baltic Sea in the north, the Eastern Front began to crumble and became the Great Retreat.[23]

Ironically, since too many generals ignored the scouting reports of their aviators, aircraft now took on a critical role in the dramatic shift eastward into traditional Russian territory. To gain some notion of the Great Retreat's extent, one only needs to follow geographically the reconstituted Russian Second Army now under General Sergei Scheidemann. Located southwest of Warsaw in April, the headquarters of the Second Army in mid-September was Molodechno, in modern-day Belarus. The army had retreated some 350 miles in 5 months. In America today, while obeying the speed limit, that would only be a non-stop, 5-hour drive in a car on an interstate highway. In 1915, however, highways did not exist, and the dirt roads, when available, became muddy and difficult to travel, if not impassable in rain. Motorized vehicles were not common, and many suffered breakdowns on rutted roadways. Soldiers walked when there was no motorized transportation; horses pulled supply wagons and wheeled artillery—when the weather permitted. The bad news from the Russian perspective was that as the Germans moved eastward they quickly and expertly repaired or reconstructed Russian rail lines to bring food, medicines, ammunition, weapons, and replacement troops to their own advancing armies.[24]

Russian aircraft played an important part during the Great Retreat in helping the armies move eastward without falling apart in defeat. Their bombs modestly disrupted the enemy's advance; scouting reports and photographs revealed the placement and size of the opponent; and observation of the enemy line enabled Russian troops to correct their artillery fire. Before year's end virtually all general officers realized the value of using aircraft in the combat environment. Early in the Great Retreat, General Scheidemann, the Russian Second Army commander, had already recognized the work and assistance supplied by the army's pilots and observers. On May 21, 1915 (O.S.), he issued an order, to be read to all his army's soldiers, expressing his praise for the role played by aviators. The document concluded: "I note with satisfaction that the aviators' efforts for the army are not lost. All my orders are quickly and thoroughly carried out. Their efforts in aerial intelligence are executed with model efficiency. Thanks to this, aerial reconnaissance remains a valuable tool to the army staff."[25]

The Russian Second Army had a special aviation detachment, called the Twenty-First Corps Air Squadron, which flew pusher-type Voisin aircraft. Each plane carried a mounted machine gun and about one hundred pounds of smaller bombs—at maximum three of the standard thirty-five-pounders. Staff-Captain Evgraf N. Kruten' commanded the squadron after completing more than ten missions a month since having joined the detachment in September 1914. The Voisin enabled Kruten' to secure the first of his seven kills as the unit tried to intercept and prevent German aircraft from gaining intelligence on the retreating army. One of his victories occurred in France on the Western Front when Kruten', who had just written a manual on air tactics, joined a group of Russian pilots who had been sent to France in 1916, where the aviators flew against German planes. (The two countries exchanged a small number of military personnel as a symbol of their military alliance.) During the opening moments of the Great Retreat the Twenty-First Corps Air Squadron also bombed and machine-gunned German aircraft at the Godzisk Aerodrome and damaged the German-controlled railroad station at Lowicz in modern-day Poland.[26]

The most important role of the squadron, however, focused on reconnaissance. One of its pilots, Petr P. Grezo, received two Order of Saint Anne medals, Third and Fourth Class, for his participation in combat missions, including excellent photographs that revealed and located reserve German units—images that suggested to Second Army generals how to plan for German pressure-points during the enemy's offensive. Meanwhile, shortly before Scheidemann's praise for the Twenty-First Corps Air Squadron, the ace of aces—Aleksandr A. Kozakov—secured the first of his twenty kills. Originally a cavalry officer, Kozakov became increasingly interested in flying and, in January 1914, transferred into the army military aviation school that was attached to the Sevastopol' center. He completed the aeronautics course and the flight-training program in Farmans and was formally designated a Military Pilot later that year. Finally, in December 1914, he was assigned to the Fourth Corps Air Squadron of the Russian First Army, which was adjacent to and just north of the Russian Second Army.[27]

Kozakov's squadron initially flew the Morane-Saulnier monoplane, powered by an 80-hp Gnome-Rhône engine. As early as December 1914, he carried a couple of altered light artillery shells that he would drop to harass German encampments during reconnaissance flights. Like Nesterov, Kozakov enjoyed fame for his unique effort to destroy an enemy plane. He packed a line with blocks of gun cotton (nitrocellulose) weighted by a boat's small anchor. When he tried the device on a German Albatros in the spring of 1915, the anchor ended up dangling underneath the enemy plane without exploding. Out of frustration, Kozakov duplicated Nesterov's *taran*, slamming his landing gear into the center of the German biplane's upper wing. The wings folded and the Albatros fell like a rock. Kozakov's landing gear collapsed into the fuselage of the Morane-Saulnier; he had to turn off his engine because his propeller had been damaged when he struck the German plane. Unlike Nesterov, Kozakov kept his safety belt on in the cockpit and he flew his aircraft like a glider, landing safely on Russian-held ground. Along with his first victory, Kozakov earned fame and honors and a reputation for high courage.[28]

While aviators such as Kozakov and Kruten' of the Russian First and Second armies took on essential duties during the Great Retreat, their naval aviation colleagues in the Baltic encountered heightened activity as well. The problem came from a decision made by German general Erich F. W. Ludendorff, who was Field Marshal Paul von Hindenburg's chief of staff for the Eastern Front. Ludendorff had an idea that was both practical and visionary: he wanted to guarantee the continued flow of the German offensive by occupying most if not all of the Imperial Russian Army's reserve divisions. In April, as the German Eleventh Army prepared to attack the Russian Third Army, Ludendorff set up a strong cavalry and infantry force to invade the Russian Empire along the Baltic coast in Courland, which today lies in Latvia. The German force became known as the Niemen Army, named for the river south of Courland. The tactic worked exactly as Ludendorff had planned. It drew sixteen reserve divisions away from the German offensive that had started in Galicia.[29]

The Russian reserve divisions operated between Riga in the north and the Kovno Fortress, which was actually south of the province. Ludendorff hoped that the German forces could take Riga as prelude to their advance against Petrograd. More than likely, a German takeover of Petrograd would have resulted in Russia's surrender to a victorious Germany, but Ludendorff was not able to capture Riga for another two years.

In Ludendorff's mind, there were two reasons for the delay—first, what he considered to be a bad decision by Berlin to reduce German forces on the Eastern Front; and second, the strong defense put up by the Russian army, navy, and naval aviation. Regardless, on May 8, 1915 (N.S.), a combined German navy and army operation took the Courland port of Libau, just north of the German post of Memel on the Baltic coast. Over time, the Germans turned the port into an important support base for their fleet and aircraft. In response, Stavka became more generous in transferring FBA flying boats to the Russian Baltic Fleet.[30]

CHAPTER 5

FLIGHT DURING THE GREAT RETREAT

★ ★ ★ ★ ★

IN 1915 THE MOST widely used aircraft in the Baltic was the FBA flying boat. Admiral Nikolai von Essen, who commanded the Baltic Fleet, proved to be both a visionary and dynamic leader. He integrated aviation with the fleet to extend its sight range beyond eight miles, the point at sea level at which the funnels and smoke of enemy ships begin to disappear due to the curvature of the earth. Over time the military developed plans to build a force of eighteen aircraft squadrons in two brigades. It never happened. Eventually nine naval air stations were established, each with a fully manned and fully equipped detachment. They dotted the coast from Revel on the Gulf of Finland south to the Gulf of Riga, with several on the islands at its mouth. A tenth detachment sailed on the seaplane carrier *Orlitsa*, with six to eight aircraft divided between forward and aft storage areas on the ship's top deck. Three of the naval air stations became extremely important—on Ösel Island at Kilkond in the west, in Arensburg in the east, and at Zerel on Sworbe Peninsula on the island's southern tip. The Irben Strait, between that peninsula and northern shore of Courland, served as the main entrance for ships steaming into the Gulf of Riga.[1]

When the naval air station on Ösel Island began operations in 1915, Russian pilots generally flew FBAs, which carried 4 bombs weighing about

35 pounds each. But German aircraft posed no threat to the Russian FBAs, even in May, when the Germans captured Libau and began to establish an aerodrome. The advanced Fokker Eindecker, which carried a synchronized machine gun and was only available on the Western Front during the summer, had a flight range of about 225 miles and simply could not make a round trip between Libau and Ösel Island. As a result, the Russians did not equip the FBAs with the field-grade Maxim 7.62-mm machine gun that would become standard in the future on Grigorovich M-5 and M-9 flying boats. FBAs, flown by a 2-man crew of pilot and observer-mechanic, conducted protective reconnaissance missions around Ösel Island and the other three islands of Dägo, Worm, and Moon. Squadrons assigned to the naval air stations were joined by the *Orlitsa*. The Russians looked for German warships and dropped bombs on enemy gunboats to prevent them from scouting the islands. After the Russian navy spread mines across the Irben Strait, the FBAs used their bombs to discourage German minesweepers from reopening the strait to German vessels.[2]

Unfortunately for Russia, less than two weeks after the Germans took Libau, Essen died from a respiratory illness that may have been pneumonia. Before long his place was taken by Vice Admiral Vasilii A. Kanin. The elderly officer chose not to disturb the pattern and use of aircraft that had been established by the time that he replaced Essen as the Baltic Fleet commander. There was an interesting peculiarity about the northern section of the Baltic Sea, however. Kanin realized that the naval war with Germany would be impossible to prosecute when the sea froze over each winter. But during the summer and fall of 1915, Russian naval aviation missions took place daily, weather permitting. Other than the occasional accidents and engine problems, basic reconnaissance flights usually posed no threat to the FBAs; undisturbed, they looked for German warships and submarines and reported their locations via radio when they returned to the naval air station. Even so, dropping relatively small bombs on German gunboats proved dangerous: the Germans carried machine guns, and although their bullets often struck the thick wooden hulls of the FBAs without causing any harm, they could damage control surfaces, put holes in fabric covering control surfaces, or hit pilots and observers, resulting in wounds or death.[3]

A case in point involved Seversky. His Russian passport—written in French, then the international language, rather than Russian Cyrillic—gave him his American name, Alexander P. de Seversky, but in Russia the military kept his official records as Aleksandr Nikolaevich Prokof'ev-Severskii. In many ways Seversky fit the standard profile for the Russian military pilots of the early days of aviation. Born into the nobility and the Russian Orthodox church, he attended military schools and the Imperial Russian Naval Academy, where he studied math and engineering. In aviation, however, Seversky had an unusual edge over most of his contemporaries. Because his father was a pioneer aviator and the owner of the empire's first factory-built aircraft (as well as having a Farman plane for his personal use), Seversky grew up around airplanes and learned to fly as a teenager. After completing an aeronautics class in Moscow taught by by Zhukovskii, Seversky went to Gatchina, where he took a flight on a Farman Model IV. His army instructor was his brother George. The reorientation flight lasted less than seven minutes; Alexander already had flown in a Farman many times with his father. He then soloed and transferred immediately to Sevastopol', where he attended both army and navy classes, flew the more advanced Voisin Model III airplane, and finished the training program in May 1915 as a junior lieutenant with the designation of Naval Pilot.[4]

Ordered north to the Baltic theater in June, Seversky spent time at the naval air station at Revel. He made familiarity flights in an FBA flying boat that had been built under license by the Lebedev Aeronautics Company of Petrograd. (Powered by an imported French Clerget rotary engine of 130 hp, the plane could barely reach the speed of 60 mph when loaded with four bombs and two crew members.) The navy then transferred Seversky to the Second Bombing-Reconnaissance Squadron, headquartered at the relatively new naval air station at Kilkond on Ösel Island. Lieutenant Vladimir A. Litvinov, the squadron leader, assigned his new pilot to a billet and to an FBA. Seversky's mechanic-observer, Sergeant Blinov, serviced the engine and installed four small bombs on racks located on each side of the cockpit. To prevent enemy bullets from striking the detonation caps, the business end of each bomb was racked upright and crowned by a cap guard. On July 6 (O.S.) two German gunboats had been spotted in a nearby cove. Since most Russian warships were farther south, the single destroyer in Kilkond Bay was under orders to stay put and prevent German bombardment of the air station facilities and planes.[5]

Lieutenant Litvinov brought news of the German gunboats to the pilots at the small Officer's Club. They were playing cards, telling war stories, and listening to phonograph disks. The squadron commander asked for a volunteer to join him in dropping bombs on the gunboats. Thirty-five pounders would not likely do much damage to German boats, but a surprise attack would force the Germans to leave their cozy anchorage. Seversky volunteered. Soon two armed FBAs from the naval station lifted smoothly off Kilkond Bay during the early evening. Even though the season of "white nights" should have enabled the two planes to fly safely together, cloudy weather made it so dark that Seversky did not see Litvinov's hand signals indicating that he was going to turn back to the air station. Seversky and Blinov flew alone to the nearby cove where the German war vessels had dropped anchor. The Germans, of course, were unpleasantly surprised that their nighttime resting-place had been discovered. Unfortunately for the Russian pilot and observer, however, the Clerget's rotary engines made such a loud clatter that the Germans had time to man their machine guns.[6]

The Russians made just one pass over the gunboats, but they managed to drop three of the four bombs. One actually hit the deck of one vessel, but there was no massive explosion; the bomb did not hit ordnance, such as mines stored on the top deck, that might actually sink or seriously damage the gunboat. During the FBA pass, however, the Germans peppered the flying boat with machine-gun bullets, many of which hit and splintered the thick wooden hulls, but others pockmarked the fabric of the wings and control surfaces. Seversky realized that his plane could not survive a second pass and headed back to the naval air station. The numerous holes in the FBA's fabric covering made the plane difficult to fly, and it lost altitude; Seversky was barely able to reach Kilkond Bay. As the plane landed roughly Blinov tried to reinstall the detonation cap on the fourth bomb, but it exploded, instantly killing Blinov and hurling the pilot into the bay. The cold water revived a dazed Seversky enough to regain the surface, where he grabbed hold of one of the two wing floats. By the time the Russian destroyer responded to the explosion and picked Seversky up, he was experiencing excruciating pain and realized that a portion of his right leg had turned into "mushy nothingness."[7]

While the number of bombing missions increased in the Baltic, the Black Sea Fleet actually carried and used more aircraft. Air reconnaissance on the sea's northern shore helped shield naval facilities and population centers. In addition, hydrocruiser task forces continued to assault fortresses around the Bosporus and to intercept coal-carrying Turkish ships. In May, the hydrocruisers *Nikolai* and *Almaz*, accompanied by an assortment of Russian warships, attacked towns and military targets near the Bosporus. When several dreadnoughts bombarded batteries, five planes from the *Nikolai* screened the ships by watching for enemy vessels. Later, aircraft from both the *Nikolai* and the *Almaz* dropped bombs on the Turkish town of Igneada. The task force then sailed east to coastal Anatolia. While aircraft spotted coal-carrying ships, the warship squadron sank seven Turkish steamers and thirty-one sailing ships and captured the steamer *Amalia*. On May 10 (N.S.) the Russians' success in disrupting the movement of coal prompted the Turko-German battlecruiser *Goeben* to try to protect Turkish transports; two hits from Russian dreadnoughts forced the *Goeben* to retreat.[8]

Through the spring and summer of 1915 Russian warships returned to the Bosporus and the Anatolia coast regularly—about a six times a month. During this period, Curtiss seaplanes and flying boats suffered both accidents from wear and tear and damage from enemy guns. Replacements consisted of FBAs built by Dukh and Grigorovich flying boats built by Shchetinin. The Grigoroviches gradually became the standard aircraft. On June 7 (N.S.), during one of the Russian assaults on Turkish ships, the destroyer *Pronzitel'nii* (piercing) stopped the steamer *Edinçik* near the town of Kozlu just west of Zonguldak. Before sinking the Turkish vessel, Russian naval officers boarded the ship and discovered documents that revealed that two German submarines, *UB-8* and *UB-21*, had recently arrived in Constantinople. The Russians had long suspected that enemy submarines had joined the *Breslau* and *Goeben*. They were correct. The two new undersea arrivals had joined the *UB-7* and *UB-14* and two small U-boats, the *UC-13* and *UC-15*.[9]

Land- and hydrocruiser-based aircraft now had the added job of watching for German submarines. The subs posed few threats to larger

Russian warships, especially those that sailed in formations that included destroyers, cruisers, or dreadnoughts; firing a torpedo would bring on a deadly Russian response. Submarines could cause problems for isolated civilian or military transports, however. For example, *UB-14* sank the steamer *Katia* on October 7 (N.S.) and the tanker *Apsheron* on October 8 (N.S.), in the northern part of the Black Sea. *UB-14* then went briefly to Varna, Bulgaria's main port, before returning to Constantinople. As early as September 30 (N.S.), Varna had become an auxiliary base for German U-boats, beginning with the small *UC-13* submarine. Earlier in September Bulgaria had secretly joined the Central Powers—a step that King Ferdinand took to expand the country's borders and overturn the results of the 1913 Second Balkan War.[10]

By September the Germans, accompanied by Austrians, seemed to have defeated the Russians during the Great Retreat by pushing them well within the tsarist empire. And with determination and perseverance, the Turks completely stalled the Allied Expeditionary invasion of Gallipoli. Moreover, General Erich von Falkenhayn, Germany's chief of staff, decided that the German victory over Russia could best be celebrated by sending General Mackensen and a force of 330,000 elite troops south to defeat pesky Serbia and its flanking threat to Austria's armies. Even before the attack on Serbia began, the large troop movement produced two significant results. First, the reduction of German soldiers on Russia's Northwestern Front ended the offensive. General Ludendorff's expectation of pushing forward and actually ending the conflict, possibly by capturing Russia's capital city, simply evaporated. Indeed, the reduced German force had to stop and dig defensive trenches, which stabilized the Northwestern Front and gave Russian armies breathing space to rebuild, rearm, and recover from the Great Retreat. The static front also facilitated the use of artillery observation balloons and, on both sides, the use of aircraft to destroy them.[11]

Second, on joining the Central Powers Bulgaria had promised to attack Serbia from the east as the Germans came in from the north. In return, Bulgaria received parts of the Balkans, especially Macedonia. The poor Macedonians, who were under Serbian control after the Second Balkan War, had to change the endings on their names from the Serbian "itch" to the Bulgarian "ov" or face imprisonment and torture. On October 6 (N.S.)

Bulgaria publicly announced that it had joined the Central Powers. It did nothing to augment the Turko-German position against Russia in the Black Sea. Bulgaria possessed only one gunboat—the ancient *Nadezhda* (hope)—and six small torpedo boats. As early as October 10 (N.S.) Russian sea squadrons began shelling locations on the Bulgarian coast. On October 27 (N.S.) a hydrocruiser task force built around the *Almaz* and the *Nikolai* shelled Varna's docks while aircraft bombed the city. The Imperial Russian Navy and its planes, however, could not stop the Bulgarian army from invading Serbia at the same time as Germany. Overwhelmed by November, remnants of the Serbian army crossed over Albania to the coast of the Adriatic Sea. British ships saved Serbian soldiers and transferred them to the island of Corfu.[12]

While Russian naval and aviation personnel held their own in the Baltic and Black Seas between May and September 1915, the Imperial Russian Army struggled to remain a cohesive opponent during the enemy offensive. The EVK helped in preserving army units from a complete collapse in the face of shortfalls in rifles and all types of ammunition. Squadron planes began to separate and fly to different battle zones in need of support from the large aircraft. In May, for example, two Il'ia Muromets planes left Iablonna for Lvov, where the disintegrating Russian Third Army was overcome by the German Eleventh and Austrian Fourth armies. While the Russian retreat continued, the EVK disrupted the enemy advance by bombing trains, depots, and headquarters and taking photographs of troop concentrations to help Russian army commanders understand the strengths and weaknesses of the German advance. The most spectacular event, on June 27 (N.S.), occurred when Lieutenant Iosif S. Bashko and his Il'ia Muromets crew bombed a German train at Przevorsk in Galicia.[13]

The plane's bombs detonated 30,000 artillery shells, which destroyed the train and damaged the railway station. Austrian POWs and a later German newspaper account confirmed the number of shells lost. The reconnaissance photos clearly revealed the extensive damage. It took the Germans several days to clean up the debris and make repairs before supply trains could return to the station. Bashko received a promotion to staff-captain on July 7 (O.S.) and the Order of Saint Anne, Second Class, with Crossed Swords, as a sign of his military bravery. He and his crew had already enjoyed a dozen successful missions since April. Like Seversky, Bashko was different from most officers who had become pilots before 1915. In his case, however, it was his background that made him distinct. He was born in Dvinsk by the Dvina River, which flows from the southeastern portion of Courland to the city of Riga and into the Gulf.[14]

More significant, Bashko was Roman Catholic in an Orthodox empire and was the offspring of ancestors who belonged to the peasant class rather than the aristocracy. At the same time, in Dvinsk he had access to a decent education through the secondary level. After graduation, he joined the army, entered the Vladimir Military School, and completed its program in 1910 with the rank of junior lieutenant. He then served briefly in the Thirteenth Sapper Battalion. In October 1911 he transferred to the Officers' Aeronautics School. He proved to be such a good flight student that a year later he was assigned to stay at Gatchina as a flight-training officer. Bashko then served in several different slots, ranging from officer in charge of the library to commander of the aviation section at the Officers' Aeronautics School. A superb pilot with many hours in the air, he possessed a great background in teaching others about flight and how to fly. From this experience, on October 8, 1914 (O.S.), Bashko logically was selected to join what became the EVK. He first studied and trained to become a deputy commander to Gorshkov, and then became commander of an Il'ia Muromets aircraft on his own.[15]

On July 19 (N.S.), after Bashko and his crew enjoyed great success in destroying enemy shells, they made a serious mistake. Since their most recent series of missions—deep behind German lines—had gone unchallenged by enemy fire, they chose to load more bombs than normal for a mission, offsetting the extra weight by removing four heavy machine guns. For their protection, their Il'ia Muromets carried just one carbine and one light Madsen rapid-firing gun. After flying well over twenty-five miles into enemy-held territory, the reconnaissance-bomber dropped nine bombs on trains, five bombs on an aerodrome, and twelve incendiary bombs on sixty German horse-drawn wagons south of the town of Shebrzheshin. Wakened before 5:30 a.m. by explosions on their aerodrome, three pairs of German pilots and machine-gun-wielding observers took off separately in their surviving LVG B.Is aircraft. The first German plane caught up with the

Russians as they left Shebrzheshin for their home base to the east. The Madsen machine gun, with a more stable firing platform, easily damaged the German plane and may have wounded one or both crew members, who successfully flew away. In the firing process, however, the Madsen recoil spring on the Il'ia Muromets gun broke, and the weapon jammed as the other two German planes approached. Defended by only a single carbine, the Il'ia Muromets took at least sixty hits from German bullets. The rounds struck oil and gasoline tanks, engine radiators, and cockpit glass. They also hit Bashko's right leg near his knee and the right side of his buttocks, and a bullet fragment and shattered cockpit glass grazed his head. The engines on the plane's starboard side stopped turning propellers.[16]

As crew members applied dressings to Bashko's leg and buttocks and cleaned and dressed his head where he had been grazed by glass and metal, the plane's deputy commander, Lieutenant Mikhail V. Smirnov, temporarily flew the Il'ia Muromets. After emergency medical care, Bashko felt well enough to resume piloting the plane. It was not an easy task. With only the two port engines still functioning, the aircraft developed a yaw to the right. But by the time the plane crossed Russian lines and reached the airfield of the Twenty-Fourth Corps Air Squadron, all four engines had stopped working and Bashko had to land the plane as a glider. The pilot made a safe enough landing and was quickly taken to a nearby hospital. He convalesced in Petrograd and returned to duty completely healed in September. All crew members received decorations. Bashko acquired the Order of Saint George, Fourth Class, after Captain Georgii Gorshkov, the EVK's senior pilot, sent a detailed report to Stavka on Bashko and his heroic crew.[17]

The field where the Russians landed was not the one used by the two Il'ia Muromets planes. Thanks to the German offensive, they had a separate field farther east after moving from Lvov to Lublin and then to Vlodava. When Bashko landed at the forward squadron aerodrome for small planes, the bomber's landing gear dragged on the damp field, fracturing several struts. The plane was repairable. In July it was ready to fly again, and it was flown to the headquarters field of the EVK, which had moved to Lida from Brest-Litovsk after leaving the squadron's original base at Iablonna, near Warsaw. Lida proved to be an excellent aerodrome for large aircraft. By then dirigibles had been removed from combat duty, and the EVK could house Il'ia Muromets aircraft in the huge empty hangars that had been built for Russian airships. Meanwhile, the second big plane at Vlodava, commanded by Lieutenant Dimitrii A. Ozerskii, continued flying combat missions to assist the remnant of the Russian Third Army, which a deputy to General Alfred Knox, the British military attaché, had described uncharitably as "a harmless mob."[18]

The British officer's comment ignored the fact that German artillery and small-arms fire had eliminated by death or capture 60 percent of the Russian Third Army's 232,000 soldiers. One can only imagine what was running through the minds of the minority who survived. Moreover, no Russian institution could miraculously produce the nonexistent bullets, shells, rifles, cannon, and 400,000 replacement troops needed to stop the German advance on the Eastern Front. Expectedly, the Russian Third Army commander, General Radko Dmitriev, had been dismissed, but instead of facing a court-martial for his incompetence in exposing his men and artillery to enemy harm, he was rewarded by being named to command the Russian Twelfth Army. Nikolai II ordered the appointment when the tsar took over the army later that summer. Such tsarist decisions begin to explain why Nikolai II lost his ruling position in 1917. Regardless, the remnant of the Russian Third Army acquired a new set of commanders, led by General Ivan Pavlovich Romanovskii, the chief of staff. Unlike Dmitriev, who ignored aviators' warnings about German forces, Romanovskii praised the EVK and the work of big planes in scouting and bombing the enemy. The EVK's information on and damage to the enemy made its contribution to the resurrection of that army under its new leadership.[19]

Ozerskii, who piloted the second Il'ia Muromets, had a career similar to Bashko's. He became a flight instructor at the Gatchina training center and in October 1914 was chosen to become a commander of one of the four-engine aircraft. But Ozerskii came from a more exalted background. He belonged to the empire's Orthodox Church, and his father was an army colonel and a member of the nobility. Before Ozerskii transferred to aviation he had served in the First Grenadier Guards, which had been founded during the reign of Tsar Aleksandr II. After July 19 (N.S.) he continued to assist the depleted Russian Third Army with his crew of three—a deputy commander, Lieutenant Mikhail P. Spasov; a reconnaissance specialist,

Lieutenant Colonel Zvegintsev, and an engine mechanic. Ozerskii and his crew conducted numerous and dangerous combat missions that enabled General Romanovskii and his staff to understand and respond to the deployment of enemy troops. For one week the Il'ia Muromets aircraft flew twice a day to bomb the railroad station and trains near Rava-Russkaia, northwest of Lvov—the terminus of the railway line that supplied Austrian units with food, ammunition, medicines, and replacements.[20]

At EVK headquarters at Lida, Sikorsky continued to modify and redesign the Il'ia Muromets based on the combat experiences of the early models flown by Bashko, Ozerskii, and other pilots. There were several Kievskii models—the name that Tsar Nikolai II gave the plane after its round trip between Petrograd and Kiev. The other four-engine planes were identified by a letter, along with the number of the plane that was built under that letter designation—i.e., the G-2 was the second plane constructed exactly the same as the G-1. The catch to this identification system was that the letters followed the order of the Cyrillic alphabet—А, Б, В, Г, Д, Е, rather than the Latin A, B, V, G, D, E. By the time the G model came out, late in 1915, the aircraft had undergone numerous changes, but two of them proved crucial to the survival of the plane and its crew. First, Germans who flew against the large Russian aircraft quickly discovered a weakness in its defense: if they approached an Il'ia Muromets plane from the rear, its crew had no way of spotting or firing a machine gun at the German plane. To overcome that deficiency, Sikorsky designed a space for a gunner-spotter in the tail of the G, D, and E models.[21]

Second, the Germans started using explosive bullets, which could harm wood and fabric airplanes in endless ways. The most obvious targets on the Il'ia Muromets planes were the exposed fuel tanks—initially four of them, one directly above each engine. When a partially filled gasoline tank was pierced by exploding bullets it most likely would erupt in fire, potentially causing fatal damage to the plane or, at minimum, starving the engine(s) of fuel. Fortunately, the Russians acquired German bullets and machine guns during the battles, partly from the occasional German plane that ended up on Russian-controlled territory because of accident, combat, or geographical error. Sikorsky used the German guns and ordnance to run experiments against discarded gasoline tanks. The results prompted the designer to change the number, location, and construction of the tanks. Instead of four tanks for the four engines, he placed two larger, more robust tanks above the center of the top wing.[22]

As Sikorsky worked on redesigning the Il'ia Muromets, the German offensive continued. In mid-September, after the enemy took Warsaw and Brest-Litovsk, it captured and occupied Baranovichi—the former headquarters of the Russian military. As a result, Stavka had to relocate further eastward at Mogilëv (perhaps appropriately, the root word in the city's name means "grave," as in a place of burial). At the very point that the Germans declared victory, sent troops to Serbia, stalled the offensive, and dug defensive trenches, Tsar Nikolai II showed up at Stavka and took personal command of the Russian military. If the army failed, it would likely contribute to the failure of the tsar as the empire's ruler. The tsar replaced his uncle, Grand Duke Nikolai Nikolaevich, who went to the Russo-Turkish Caucasian Front, and also appointed General Mikhail Alekseev to succeed Chief of Staff Nikolai Ianushkevich. The new, self-effacing chief of staff proved to be a genius in military organization. And he took advantage of the fact the Germans had stalled their offensive, using the relative peace to revive, augment, supply, and rebuild the Russian army on the Eastern Front.[23]

As these events unfolded, the German offensive stalled west of Minsk, the modern-day capital of Belarus, forcing the EVK to leave Lida for Pskov, which became its new headquarters. It was not far from Petrograd, where the big planes were produced, and the Russian Twelfth Army protected the area, including Riga, from the German Niemen Army. During the winter of 1915–1916 the Russians decided to continue to break up the larger squadron into several smaller detachments in order to assist Russian armies in key locations on the Eastern Front. Meanwhile, Ozerskii and his crew also had to move, in their case from Vlodava eastward to Slutsk. It was a safe area close to fifty miles south of Minsk. The Il'ia Muromets crew helped the Russian Third and Fourth armies by conducting deep reconnaissance and bombing runs. On a mission on November 15 (N.S.), the plane unloaded eight bombs on the Baranovichi railway station. Because the station was a key supply depot, the Germans had protected the facility with antiaircraft guns; exploding shells had damaged cables leading to the ailerons. Using the rudder, Ozerskii successfully redirected the plane to his aerodrome at Slutsk. It is not known for certain, but perhaps a gust of wind caused the aircraft's wings to lose their parallel attitude to the ground. Without ailerons the attitude of the plane could not be corrected to maintain lift, so the craft slipped rapidly and crashed, killing the entire crew except for Lieutenant Spasov, the deputy commander, who survived with multiple injuries and bone fractures.[24]

A month before this tragedy, the aircraft at the EVK headquarters at Zegevol'd Aerodrome near Pskov had been active in photographing and bombing the Germans' static position. For example, on October 18, 1915 (N.S.), three Il'ia Muromets aircraft, including one piloted by Staff-Captain Robert L. Nizhevskii, flew south of Riga in Courland territory occupied by the German Niemen Army. While Nizhevskii possessed many of the same attributes as other early pilots, he came from a different social and service background. Born in the Riazan Province southeast of Moscow in 1885, at age thirty he was several years older than most pilots. Although his ancestry made him part of the aristocracy, he attended the Roman Catholic Church. The staff-captain completed his education at military schools, but after a brief tour with the Twenty-First Sapper Battalion, he was assigned to a balloon detachment. Logically, he then transferred to the Aeronautical Training Park at Volkhov for balloon aeronauts and later became a prominent member of the institution's staff.[25]

Many aircraft pilots had first flown as aeronauts in balloons. Nizhevskii, however, was a pioneer in Russia's airship program; he studied the vehicle in France and then commanded the dirigible *Astra* (aster) in Russia. When the military finally realized that the flammable gas that buoyed the airship made it an easy victim of enemy gunfire, Stavka decided that dirigibles were unsuitable for combat. At the time, generals over ground forces assumed that dirigible commanders could be easily transferred to the EVK. General Shidlovskii discouraged that idea on grounds that airships and aircraft flew differently in terms of speed, controls, and changes in attitude and performed differently in takeoffs and landings. In the summer of 1915, Nizhevskii may have been the only dirigible commander who succeeded in becoming an EVK pilot. Several months later he joined lieutenants Georgii V. Alekhnovich and Aleksei V. Pankrat'ev in bombing the railway station near the predominantly Jewish town of Friedrichstadt, close to the Düne River in Courland; the station received supplies and ammunition for the German Niemen Army. While two of the planes dropped dozens of 35-pound bombs, the Il'ia Muromets aircraft piloted by Nizhevskii carried a single 540-pound bomb, which destroyed 2 storage buildings. This would not be the last time the EVK would visit the Friedrichstadt station. By the end of 1915, the EVK had completed more than 100 missions, dropped 20 tons of bombs, and supplied army field commanders with hundreds of high-quality photographs of enemy positions.[26]

There is a significant contrast between the record of the big four-engine reconnaissance-bombers and that of small single-engine planes in 1915. That year the EVK lost one airplane and three men after a combat mission. By contrast, the casualty rate for pilots of the smaller craft reached 30 percent by December; the percentage was even higher for the monthly loss-rate of aircraft. Death and injury occurred not only because of enemy fire but in equal measure from inexperience and poorly constructed planes. This is why both aircraft production in Russia and airplane purchases abroad had to increase. It also explains why the establishment of pilot-training programs in various locations across the empire was the only way to overcome the critical shortage of pilots. The pressure for more pilots forced training centers to cut corners, to the detriment of novice pilots who were flying in combat zones. As far as one can speak of air superiority in 1915, the Germans possessed it because they generally had better pilots who flew better planes.[27] In truth, full-fledged air superiority on the Eastern Front would not come until 1916, when real fighter aircraft began to appear.

Nevertheless, Russia had a number of successful combat flyers of single-engine planes. Three interesting examples are Konstantin K. Vakulovskii, Ivan A. Orlov, and Petr-Eduard M. Tomson. Like most of their contemporaries who flew non-fighter missions in 1915, none of them shot down a single enemy plane that year; as a result, they provide excellent examples of how most army pilots assisted ground forces during the war. Unlike many others, however, by 1917 two of them—Vakulovskii and Orlov—had become ace pilots. Tomson never attained that status. Although he did shoot down several planes, he only received confirmation for destroying one enemy plane in 1917. Despite this, he earned many awards for his outstanding work and bravery, especially in his flights under heavy enemy fire.²⁸

Second Lieutenant (*podporuchik*) Vakulovskii earned the Order of Saint George, Fourth Class, for his flight under German small-arms fire in 1915. A major general's son who belonged to the empire's favored church, he completed a military education and graduated from the Nikolaevskii Engineering School in 1914. On August 8 (O.S.) of that year, Vakulovskii became a staff observer of the aviation detachment at the Novogeorgievsk Fortress, located about twenty miles northwest of Warsaw, by the Vistula River. The young lieutenant became enamored of flight and took the opportunity to attend the nearby flight-training center at Warsaw. Shortly before the German offensive in the spring of 1915, Vakulovskii joined the Novogeorgievsk Air Squadron as a pilot. There was a problem with the fortress, however. The nineteenth-century bastion contained more than 1,600 artillery pieces and almost a million rounds of shells, but it was built of brick, not concrete, and was vulnerable to explosive ordnance fired by field guns. By August the Germans had surrounded the structure with heavy artillery and turned much of the building into rubble. Shortly before the Russians finally surrendered, Vakulovskii was able to fly east to safety with the fortress' standards and sacred cross, but only after an immensely dangerous takeoff that drew withering fire from German troops.²⁹

Ivan Orlov's path to aviation was quite different from Vakulovskii's. Born into a wealthy family of nobility, he enjoyed a strong non-military education and studied law at the university level. His attendance at a prewar airplane festival near his home in Saint Petersburg piqued his interest in flight. With family support, Orlov built his own plane, joined the Imperial All-Russian Aero Club, and secured a pilot's license in June 1914. When war broke out, he volunteered his services—much like some other civilian pilots who belonged to the empire's Aero Club. Although he entered the army as a private (*riadovoi*), he soon moved up to private first class (*efreitor*), then junior noncommissioned officer (*mladshii unter-ofitser*), and finally to ensign (*praporshchik*), at the beginning of 1915. By then he had been decorated four times for his outstanding reconnaissance work and given special flight training to qualify him to fly Voisin aircraft. Transferred to the Russian First Army and the Thirty-Seventh Corps Air Squadron, Orlov presented excellent scouting reports and photographs to commanders from his many combat flights. In August he would earn another decoration, the Order of Saint George, Fourth Class, for two flights he made not far from Friedrichstadt. Orlov's Voisin took a number of hits from German gunfire from a large enemy reserve force partially hidden in a wooded area. The information that he brought back enabled the Russian First Army to take steps and move forces that prevented a German breakthrough in that sector of the Russian Northwestern Front.[30]

Although Tomson never became an ace pilot, he was a remarkable aviator. His humble background in a small district near the Gulf of Riga had kept him out of any formal school, though he did learn to read and write. Moreover, his Lutheran faith would have alienated him from Grand Duke Aleksandr. Because of the strong German presence in Russia's Baltic territories, it should come as no surprise that Tomson took flying lessons and became a licensed pilot in Germany in 1912. In fact, he was in Germany when war began and escaped imprisonment by fleeing to France, where he joined the French Airplane Service. With French permission, he transferred to Russia in the spring of 1915 and became a private and pilot in the First Corps Air Squadron of the revived Russian Second Army. In July and August he conducted ten reconnaissance flights around Warsaw, which the German offensive had taken. Because Tomson was under significant German small-arms fire during his flights, his bravery and scouting success earned him the Soldier's Cross of Saint George, Fourth Class. By then he had been promoted to senior noncommissioned officer (starshii unter-ofitser). His fearless survey of enemy positions continued, and by 1917 he had received additional decorations and had been elevated to junior lieutenant.[31]

CHAPTER 6

THE HEIGHT OF THE AIR WAR

★ ★ ★ ★ ★

BEFORE EXPLORING AVIATION'S contribution to the Russian battlefield in 1916, it is important to consider some of the differences between the Eastern and Western fronts. The Western Front was much shorter, only about 400 miles, yet its combat zone contained a significant concentration of British, French, and German soldiers, accompanied by a massive number of artillery and machine guns that put in harm's way almost every square yard of the elaborate trench system that ran from the North Sea southeastward to neutral Switzerland. At the same time, this much smaller front was home to far more aircraft than were assigned to the Eastern Front. In 1916 the French and Germans each maintained an inventory of planes that approached 1,500. During the Battle of the Somme, the British employed not just 4 air detachments, as the Russians might have done, but more than 27 squadrons, equipped with 410 aircraft. In short, there was a big difference between the air activity on the Eastern and Western fronts. The Germans, for example, claimed 7,067 air combat victories in the West, but they reported only 358 triumphs in the East against Russian pilots. In the West, when weather permitted, hundreds of air missions and a dozen dogfights occurred every day, with bombers, scouting craft, and fighters on both sides. Meanwhile, in the much more expansive Eastern Front, which extended from Riga in the north to Czernowitz (near Romania) in the south, there were some minimally contested areas that never drew a single airplane from either side.[1]

The amazing thing about the Eastern Front is that by March 1916 Chief of Staff Mikhail V. Alekseev succeeded in rejuvenating twelve Russian armies. Facilitating his rebuilding effort was the Russian industrial expansion, which led to the manufacture of military hardware that put the Russian armies on a more competitive footing with enemy forces. But three critical factors enhanced the larger picture for the Russians. First, German and Austrian troops killed or captured 2.5 million Russian soldiers, whom Petrograd had to replace by enlarging the draft. Such horrific losses deeply undermined the morale of Russian troops, especially those who faced German armies in the north. Unsurprisingly, offensives on the Russian Northwestern Front faltered in 1916. Second, industry's massive effort to replenish war materiel caused disastrous problems for the economy. Russian cities could not barter machine guns for grain and other agricultural products. Collapse of the urban-rural exchange system, coupled with a diminished supply of fuel through industrial use, ignited the Russian Revolution in 1917. Third, Russia depended on foreign imports for the war—and not just airplanes and aircraft engines. Between 1914 and 1917, for example, Russia imported 836 million dollars' worth of products from the United States—then a fortune—including airplanes, aircraft engines, armored cars, barbed wire, boots, copper, cotton, dyes, electric machinery, gunpowder, harnesses, horseshoes, howitzer shells, lead, leather, locomotives, machine tools, medicines, nickel, rails, railroad cars, rifles, rubber, saddles, shrapnel, surgical instruments, trucks, wool, and zinc. Together, these imports and Russian industrial output helped create an army in 1916 that was better equipped than at any other time during the war.[2]

In goods directly related to aviation, Russian industry built over six times more aircraft in January 1916 than it had in August 1914. The original monthly production for each of the 5 major Russian firms was 25 planes for Russo-Baltic; 35 for Lebedev; 40 for Anatra; 50 for Shchetinin; and 60 for Dukh. Six smaller workshops constructed and added 6 to 8 airframes to the monthly base of 217. As might be expected, the number of aviation laborers, ranging from those assembling airframes to those building propellers, more than doubled in 1916 to 5,029 workers. Even so, the shortage of engines continued; the French motor subsidiaries in Russia did not come close to manufacturing enough aircraft engines to match the number of airframes being produced. By year's end France, Great Britain, Italy, and the United States had exported to Russia via Arkhangel'sk or Vladivostok some 2,500 aviation engines, which greatly (but not completely) helped meet the empire's requirements. By year's end these countries also exported about 900 assembled aircraft to Russia. Despite these imports and Russia's increased domestic production, the problem of supplying the Russian military with enough aircraft remained because monthly loss rates often approached 50 percent. Regardless, each of the 12 armies fielded several squadrons of planes, and heavy combat areas received additional special bombing and fighter aircraft detachments.[3]

The Russian air forces and those of several other belligerents as well appeared to reach a degree of autonomy during that period. Before war's end, Great Britain, for example, combined its Royal Flying Corps and Royal Naval Air Service to form the Royal Air Force (RAF), under Major General Hugh Trenchard as chief of staff. Theoretically, the RAF held the same status as the British army and navy. By mid-1915, Grand Duke Aleksandr commanded the Directorate of the Military Aerial Fleet. On November 24, 1916, with Order No. 1632 from the chief of the General Staff, the grand duke also took complete charge of all Air Force inspectors. In the case of Russia, however, true autonomy for aviation existed more on paper than in reality. In effect, aircraft operations continued to be administered by army field generals and navy sea commanders. (As mentioned previously, pilots could not be promoted above the rank of army captain. The lone exception was in the EVK.) When the Great Retreat forced Stavka to evacuate Baranovichi for Mogilëv, the grand duke shifted his headquarters to Kiev. Naturally he kept in touch with Stavka, but his autobiography reveals that he increasingly spent time interacting with the Romanov family and expressing concern about the tsar and the tsar's future.[4]

The grand duke admitted that he was in Petrograd quite frequently; he could justify the visits there, since three large aircraft manufacturing firms operated in the empire's capital city. He remembered: "Each time I came back to Kieff [Kiev] with my strength sapped and my mind poisoned."

From his vantage point the rumors that were spreading across the capital city about Tsar Nikolai II and his wife, Aleksandra Feodorovich, were downright ugly and wicked. The grand duke noted false rumors alleging that the tsar had become an alcoholic; that he took Mongolian drugs that clogged his brain; that his prime minister was in league with German agents in neutral Sweden; that his German-born wife favored Russia's defeat; and that she had had sexual relations with the uncouth and self-proclaimed holy man, Grigorii E. Rasputin. Rasputin's unsavory reputation, though, was very real and well-deserved. He argued that salvation followed sin, and he sinned incessantly. He exerted influence over the royal couple—including their appointment of some of his "favorites" to high ministerial positions—because his calming powers had saved the couple's only son and heir to the throne from potentially fatal episodes of severe bleeding due to hemophilia. Later that year Grand Duke Aleksandr said he had "felt glad to be rid of Rasputin" when the "holy man" was murdered after dinner in the home of Aleksandr's daughter, Irina, and son-in-law, Prince Feliks Iusupov. But Rasputin's death had come way too late to help save the Romanov dynasty.[5]

The grand duke's effectiveness as head of the Directorate of the Military Aerial Fleet was limited by his focus on Rasputin and the problems of the tsar. At the beginning of 1916, the Russian air force had reached the serious level of 53 squadrons—42 army corps, 8 fighter corps, and 3 special detachments to protect Imperial residences. The grand duke's distance from Stavka's headquarters and his uneven attention to aviation issues prevented the directorate from moving toward greater autonomy for military aircraft. And even if the grand duke had wanted to create a semi-independent air force, Stavka continued to prevent him from elevating the rank of pilots above the level of captain. Fortunately, the grand duke had nothing to do with the EVK, which he disliked; it remained in the hands of Stavka. The Il'ia Muromets aircraft, formidable weapons for their time, conducted 442 combat missions during the war. The reconnaissance-bombers destroyed 40 enemy planes, took 7,000 high-quality photographs of enemy positions, and dropped more than 2,000 bombs. By the end of 1916, their varied engines could be rated as high as 225 hp; they carried some armor plating, were equipped with protected fuel tanks, and housed 7 or more machine guns and occasionally a small-bore, quick-firing cannon. The plane represented an early version of what Americans would later call the B-17 flying fortress.[6]

Once the EVK headquarters had been established at Zegevol'd, close to Pskov but not far from Riga in the north, Russian authorities created two other detachments for the 1916 campaign that were close to the middle and south of the Eastern Front—Stan'kovo, near Minsk, and Kolodziievska, near Tarnopol in the region of Galicia. It is easy to argue that Zegevol'd was the most important aerodrome for an EVK detachment. If the Germans were to capture Riga, the enemy would inherit a direct rail line to the empire's capital in Petrograd. Riga's extremely important strategic position helps explain why the EVK headquarters was stationed near that city. Understandably, Stavka wanted to do everything necessary to keep the Germans from taking Riga, which seemingly was the objective most desired by General Ludendorff, who planned the German campaign along the Baltic shore. Nevertheless, by the end of the German offensive enemy troops ended up stalled and dug in about twenty-five miles from that key urban center. During the later winter of 1916 when weather permitted the EVK detachment at Zegevol'd began regular photographic reconnaissance in Courland and the southwestern shore of the Gulf of Riga detailing the German positions for the Russian Twelfth and Fifth armies.[7]

In April 1916 the photographs reinforced the need for the EVK detachment at Zegevol'd to repeat its bombing of the railroad station in Courland at Friedrichstadt, which supplied the Niemen Army. The EVK sent the tenth version of the Il'ia Muromets Model V, piloted by Lieutenant Avenir M. Konstenchik. Born the son of a Russian Orthodox priest in the city of Grodno (now in Belarus), Konstenchik completed his basic education by attending a Russian Orthodox seminary. Yet the young man chose not to continue his seminary studies, which would have led to priesthood, but rather attended a special military academy. After graduation, he served in the Thirty-Third Infantry Regiment, then transferred to aviation and became a pilot. His first aviation assignment took him to the air squadron attached to the Brest-Litovsk Fortress. In September 1914 Konstenchik was selected to be one of the first pilots to learn how to fly one of the ten Il'ia Muromets aircraft ordered by the military.[8]

Those joining Konstenchik on the April 13 (O.S.) mission included the deputy commander, Lieutenant Viktor F. Iankovius; artillery officer and reconnaissance photographer, Lieutenant Georgii N. Shmeur; machine gunner, Sergeant Major Vladimir Kasatkin; and engine mechanic, Sergeant Major Marcel Pliat. Pliat had been assigned under an exchange program with France designed to help cement the military alliance between the two countries and enable them to share air combat techniques. Besides maintaining and repairing engines, Pliat manned the tail machine gun. He also was a special EVK member because of his Franco-African background. He proved to be the savior of the plane and crew in keeping a couple of Sunbeam motors in operation during flight. Imported from Great Britain, each Sunbeam liquid-cooled engine produced a rating of only 150 hp. Fortuitously, Sikorsky's redesign of the big plane resulted in a craft that was so aerodynamically improved that it could carry crew and bombs to an altitude above 7,800 feet.[9]

Being able to fly at higher altitudes kept the reconnaissance-bomber well above small-arms ground fire from the enemy. The first pass over the large Daudzevas railroad station near Friedrichstadt went smoothly enough as the Il'ia Muromets dropped a half dozen of its thirteen bombs. During the return pass to finish the bombing run and take photographs, the Russians discovered how much the Germans valued the huge station and its warehouses. The enemy protected the rail facilities with guns that fired exploding ordnance that could reach high-altitude aircraft flying at 15,000 feet. A shell exploded near the cockpit; metal fragments from the shell-shattered glass wounded Konstenchik and damaged three of the four engines. Shrapnel also struck the hands of the artillery officer and shattered the camera that he was holding. When the pilot was hit with pieces of metal, he fell from his seat and pulled the steering column backward. The result was an abrupt climb, which caused a stall. As the plane dropped toward the ground, Iankovius slid into the pilot's seat. He was able to get the plane out of the stall and stabilize flight at about 3,000 feet as Kasatkin applied emergency first aid to Konstenchik. Fortunately, Pliat, in the plane's tail, held on tightly to save himself from falling away from the plane as it dropped during the stall; when the aircraft recovered from its near-fatal descent, he worked his way forward through the fuselage to join the other crew members.[10]

Pliat then climbed out on the wing and, ignoring grave danger to himself, managed to keep 2 of the plane's 4 engines running. Limited power forced the Il'ia Muromets to continue flying only at a low altitude of 1,000 meters [about 3,075 feet]. For 26 kilometers (a little more than 16 miles) the plane flew over German troops, who took pleasure in firing rifles at the Russian plane, wounding crew members. When the bullet-riddled aircraft landed at a Russian aerodrome and came to a stop, its right wing dropped completely to the ground. Despite such obvious combat damage, the plane's safe landing only added to the legend of the miracle survivability of the Il'ia Muromets. This mission resulted in a series of awards: Lieutenant Konstenchik received the Order of Saint George, Fourth Degree; Lieutenant Iankovius, the Sword of Saint George; and Sergeant Major Pliat, the Cross of Saint George. Sergeant Major Kasatkin received a commission as an officer.[11]

Meanwhile, several of large EVK planes continued to aid the Russian Twelfth and Fifth armies in their defense of Riga by photographing and bombing German targets of opportunity. Then, in July 1916, came several extraordinary events that would have a major impact on the EVK. First, future ace pilot Alexander Seversky returned to combat duty—minus a major portion of his right leg—as part of Second Bombing-Reconnaissance Squadron. Although the squadron was headquartered at the naval air station at Zerel on the southern point of Ösel Island's Sworbe Peninsula, it established an auxiliary base on tiny Runo Island, about forty-five miles east-southeast of Zerel, near the center of the Gulf of Riga. With Runo Island as a base, Russian pilots searched intensely for German submarines. In one of Seversky's first scouting flights from Runo, in a Grigorovich M-9 flying boat, he shot down a German Albatros C.Ia. The German land plane had been converted to a seaplane by replacing its wheels with floats. The victory was a jolt to Seversky, his squadron, and the Baltic Fleet. It meant that the Germans had some type of seaplane base near the Gulf of Riga. Subsequently, another squadron member in an M-9 spotted an Albatros C.Ia seaplane taking off from Lake Angern. Less than a mile into the interior, the lake parallels the western shore of the gulf for ten miles. The Germans loved it because their seaplane base was completely invisible to vessels of the gulf's Imperial Russian Navy.[12]

The German seaplane base on Lake Angern posed a threat not only to the aviation presence on Runo Island but also to the passage of Russian naval vessels in the gulf. The Russians responded quickly. Using the radio unit on Runo, the navy ordered the Second Bombing-Reconnaissance Squadron to send several bomb-loaded M-9s over the enemy seaplane station to see how much damage they could inflict. Joining Seversky for the perilous honor of attacking the Germans in August were lieutenants Diterikhs and Steklov, each of them accompanied on the right side of the M-9's cockpit by a sergeant mechanic-observer who manned a single rotating machine gun. As in the case of the FBAs of the previous year, bombs were installed left and right of the hull and next to the pilot and sergeant. The flight from Runo to the gulf shoreline adjacent to Lake Angern went smoothly enough. As the three flying boats approached the enemy's air base, however, their noisy, clattering Salmson engines warned the Germans of their arrival. Russian pilots and crews ignored the ground fire from small arms and at least one antiaircraft cannon as the three M-9 pilots focused on dropping ordnance on several station sheds.[13]

The end of the successful pass over target resulted in gains and setbacks for the Russians. The good news was that although a number of German bullets and shell fragments had struck the M-9s, the aircrafts' thick planked hulls had served as protective cocoons for pilots and sergeants; no metal fragments punctured the flesh of crew members. The bad news contained multiple parts. Lieutenant Steklov had to abandon the small detachment and flee to the northeast; an exploding shell had pocked-marked a radiator that then spewed out steam. Steklov would be saved. The M-9 carried two elongated radiators, one on each side of the liquid-cooled engine. The second intact radiator bought Steklov a few precious minutes of flight before the overheated engine froze and turned his flying boat into an unpowered glider. Fortunately for him, his motor enabled him to reach the gulf, where he was able to glide far enough from shore to be picked up safely by a Russian gunboat. Meanwhile, Diterikhs and Seversky circled back toward the enemy, intending to use their M-9 machine guns to damage the unmanned German seaplanes. But the pilots and sergeants soon discovered that their combat mission had actually just begun. Russian planes were entering a new and very dangerous phase of battle.[14]

When the M-9s approached the German base, some seaplanes already had taken off to confront the Russians. Soon Diterikhs and Seversky faced a flight of seven Albatros planes with crews eager to seek revenge for the damage caused by Russian bombs. Thus began one of the great, epic air battles of the Eastern Front, which continued for at least an hour. It was prolonged in part because of the machine gun setups on the two opposing types of aircraft. With motor and propeller in front, the Albatros biplane was pulled through the air. The pilot sat behind the engine and behind the pilot sat the observer, who operated the machine gun from the second cockpit. But the only clear firing view for the German machine gun was in the arc between the two open sides of the seaplane. Firing straight ahead would kill the pilot, and even a slightly slanted forward aim would potentially destroy the struts and wire braces that supported the wings. By contrast, the M-9 crew sat in front of the engine and propeller, which pushed the craft through the air. As a result, the M-9 machine gun swiveled and the crew had a clear shot at anything in the 180-degree arc from side to side in front of it.[15]

Besides being badly outnumbered by enemy planes, the Russians faced another problem. Due to the high drag produced by the hull, two radiators, and the Salmson engine, the M-9 had a top speed of only 110 kmh (about 68 mph). The Albatros C.Ia, with wheels for landing gear, flew 142 kmh (about 88 mph), but substituting pontoons for wheels, as the designer did, reduced the plane's speed. Nevertheless, German seaplanes remained somewhat faster than Russian flying boats. Diterikhs and Seversky understood only too well that the odds were against them in guns and speed; a frontal attack against multiple German aircraft would be suicidal. Instead the Russian pilots initiated a flight of movement and maneuver that appeared to be an aerobatic dance in the sky. The pair wove their flying boats synchronously, in and out, with tight crossovers, creating an imaginary chain that proved difficult for the Germans to penetrate. Without the benefit of radio communications, German attacks lacked any type of logical coordination. Individual Albatros aircraft moved forward and back, exchanging gunfire with the M-9s. Seversky's plane took more than thirty hits in the running gun battle, but the M-9 hull preserved its occupants, and all control surfaces remained functional.[16]

The Germans were not so fortunate. Their designers had used wooden veneer as the covering for the Albatros fuselage. Bullets penetrated the covering and exposed the crew to deadly fire. As the aerobatic chain moved the air duel away from the enemy base and into the gulf toward Runo Island, Seversky's M-9 shot down two of the German planes. All seemed to go well for the Russians until Diterikhs' machine gun jammed, leaving the Grigorovich defenseless. When an Albatros moved forward to finish off the M-9, Seversky abruptly and fearlessly changed the heading of his plane to a collision course with the Albatros, but his object was not to initiate the *taran*. Instead, his machine gun opened fire, smashing bullets into the fore and aft cockpits, killing the crew and sending the Albatros into the Gulf of Riga. Seversky's brazen action and the appearance of several more M-9s from Runo helped remind German pilots that their planes were low on fuel and they needed to return to their base at Lake Angern.[17]

As might be expected, the one-legged aviator gained promotion to senior lieutenant (*starshii leitenant*). Tsar Nikolai II, in his role as commander in chief (*glavnokomanduiushchii*) of the Imperial Russian Military, awarded Seversky the Gold Sword as Knight of the Order of Saint George. It was Imperial Russia's highest decoration and is sometimes equated with the top U.S. military decoration, the Medal of Honor. Seversky certainly demonstrated courage worthy of the decoration. Subsequent scouting reports, however, revealed that the Germans on Lake Angern quickly recovered from the M-9 bombs and the loss of the three Albatros aircraft. The enemy seaplane base posed a threat to Russian naval vessels, especially in providing air intelligence reports about Russian ships to German submarines in the Gulf of Riga. Regardless, the Baltic Fleet recommended that Stavka ask the EVK to conduct a significant bombing operation against the German seaplane station. The first thing the EVK did was to send a single Il'ia Muromets on a reconnaissance flight to the enemy base.[18]

The plane's commander, Lieutenant Vladimir Lobov, and his crew enjoyed a very successful intelligence run over the German air base. Sharp photographs clearly outlined sheds, hangars, barracks, and other facilities. A half-dozen Albatros aircraft took off and tried to intercept the Russian four-engine plane, but the bomber's crew fired so many Lewis machine guns that the German aerial attack failed completely. On September 4, 1916

(N.S.), four Il'ia Muromets aircraft left the Zegevol'd Aerodrome near Pskov under the command of Lieutenant Georgii I. Lavrov. The detachment flew to Lake Angern and dropped seventy-three bombs on the German station, which housed seventeen seaplanes. Observers on the Russian aircraft confirmed the destruction and fires that consumed aircraft, hangars, and various structures. Multiple machine guns on the four reconnaissance-bombers suppressed enemy antiaircraft fire from the ground. As the Russians headed back to their aerodrome, they noted that columns of smoke were rising where the German seaplane station had been. The four large planes suffered no damage and returned safely from their mission to Zegevol'd.[19]

The Il'ia Muromets aircraft assigned to the Stan'kovo Aerodrome near Minsk were equipped with less powerful, British-made Sunbeam engines that had a tendency to reduce the performance of the large reconnaissance-bombers. There was one IM Kievskii model that carried better Argus motors. In the summer and fall of 1916, the detachment carried out numerous bombing and photographing missions on behalf of the Russian Second, Tenth, and Fourth armies. With EVK aid these armies close to the center of the Eastern Front maintained a fairly stable combat line against the enemy. In order to distract the Germans in the north from a planned Russian offensive, the EVK decided to put on a show of force by sponsoring a major air attack against the headquarters of a German reserve division near the town of Boruna, just below the Russian offensive. The attacking force comprised four Il'ia Muromets planes and sixteen Morane-Saulnier French fighters, built by the Russian Dukh Company. The planes took off separately on September 25, 1916 (N.S.). Unfortunately, both the plans and their execution failed. The fighters missed linking up with the bombers and three of the larger aircraft never reached the target. One of the three Il'ia Muromets planes encountered German fighters supplied with explosive ordnance. An intercepted radio message later revealed that the Germans had lost three of their planes in the air battle; however, enemy bullets exploded one of the Russian bombers' fuel tanks. The plane crashed, killing the entire crew, including its commander, Lieutenant Dimitrii K. Makhsheiev. Only the IM Kievskii completed the mission in triumph; overall, the show of airpower miscarried miserably.[20]

Meanwhile, the offensive against the German forces on the Russian Northwestern Front simply failed. Occasionally, the Russian armies, under the leadership of General Aleksei N. Kuropatkin, nudged enemy troops back a short distance. Kuropatkin, the officer corps, and conscripted soldiers lost heart over Russia's ability to defeat Germany. The Kuropatkin debacle was in sharp contrast to the performance of the four armies on the Southwestern Front commanded by General Aleksei A. Brusilov. After a crushing artillery barrage on June 4, 1916 (N.S.), Brusilov's forces successfully attacked soldiers of the Austro-Hungarian Empire. A major ingredient in the spectacular advance of the Brusilov offensive involved the EVK squadron that operated close to the Russian Seventh Army, which had spearheaded the attack against enemy troops. Led by Staff-Captain Aleksei V. Pankrat'ev, the EVK detachment photographed and secured important intelligence on the disposition of Austrian units and artillery. Performing two missions a day, the large planes also bombed railway stations, railroad beds, warehouses, and towns occupied by enemy soldiers. When the Russians occupied new territory, ground troops saw first-hand evidence of the destruction caused by Il'ia Muromets aircraft and heard tales of how Austrian troops abandoned in panic their positions after a Russian bombing run.[21]

In action by single-engine planes, it should come as no surprise that Russia's top two ace pilots flew in the very active Southwestern Front, where aviation and aviators were held in high esteem by Brusilov, the front's offensive-minded commander. As noted earlier, in 1915 a type of Russian fighter aircraft emerged that employed a machine gun in the front of a pusher-type aircraft. A Grigorovich M-5 flying boat of 1915 also could carry that weapon, and in 1915–1916 a machine gun could be placed on the top wing of a Nieuport biplane that was pulled through the air by its propeller. The formal creation of fighter detachments with six aircraft each took place in March 1916 under Order No. 30, signed by Tsar Nikolai II. By August of that year, there were a dozen fighter squadrons, one for each of the twelve Russian armies. The use of such detachments could not begin to protect all aircraft that performed reconnaissance missions over enemy forces; nevertheless, on the Russian Southwestern Front, the country's second-highest ace pilot, Vasili I. Ianchenko, was credited over time with sixteen enemy kills.[22]

As a youth, Ianchenko studied mathematics and mechanical engineering at the Saratov Technical School, graduating in 1913 at age nineteen. His enthusiasm for airplanes prompted him to take flying lessons, and when war broke out, in 1914, he volunteered for military aviation service. Interestingly, because he had entered the army from the lower-middle class, failed to attend a military school, and had begun army life as a private (*riadovoi*), he never attained a rank higher than ensign (*praporshchik*). Over the winter of 1914–1915 he attended ground school at the Saint Petersburg Polytechnic Institute. Once he completed the aeronautics course he received orders to travel by train south to the Sevastopol' Aviation School. In the spring and summer of 1915, he finished military flight training there on the French-designed, Russian-built Morane-Saulnier monoplane. That fall he transferred to the Twelfth Air Corps Squadron, where he demonstrated such outstanding skills as a pilot that he received orders to attend an air school in Moscow where he learned to fly an advanced Morane-Saulnier fighter. After finishing that course early in 1916, he went to a squadron in the central regions of the Eastern Front. Ianchenko flew ten combat missions, but after hearing about the tsar's order he requested transfer to Russia's first formal fighter detachment—the Seventh Fighter Squadron, attached to the Russian Seventh Army. He joined what proved to be a busy squadron, preparing for the Brusilov offensive. Between April and October 1916, he flew eighty combat missions and became one of Russia's most decorated pilots.[23]

Russia's top ace pilot was Aleksandr A. Kozakov, who had amassed twenty confirmed kills. Somewhat older than other pilots, he was born in 1889, the son of a nobleman. He attended military schools in his youth and graduated from the Elizavetgrad Cavalry School. The junior lieutenant (*kornet*—cavalry rank) spent his first years in a horse regiment, but transferred to aviation as a senior lieutenant (*poruchik*—cavalry rank) in 1914. After completing ground school and flight training in October, he was assigned to the Fourth Corps Air Squadron, north of Warsaw, where he flew the two-seat Morane-Saulnier monowing reconnaissance plane. Near the end of the 1915 Great Retreat he was promoted to staff-captain (*shtabs-rotmistr*—cavalry rank) and appointed to head the Russian Eighth Army's Nineteenth Corps Air Squadron on the Southwestern Front. Early in 1916, on his own initiative, Kozakov had a Maxim machine gun installed on the top wing of his Nieuport 10 biplane. After multiple kills, he became the leader of the three Russian Eighth Army Corps aviation detachments that formed the First Combat Air Group. To protect the Eighth Army's recent victory in the skies over the Austrian city of Lutsk, a major railway hub, the combat group received special Nieuport 11 and SPAD SA.2 fighters imported from France.[24]

Austria hoped to regain Lutsk and destroy or damage railway facilities, so Kozakov and the First Combat Air Group engaged numerous Austrian fighter and scouting aircraft. The Austrian Brandenburg plane actually had been designed by a German, Ernst Heinkel, and originally was built in Germany by the Hansa und Brandenburgische Flugzeug-Werke. Moreover, Germans often piloted the "Austrian" airplanes. The Russians' effort succeeded. In 4 months, they captured 417,000 Austrian prisoners, 1,795 machine guns, 581 artillery pieces, and 25,000 square kilometers (15,500 square miles) of territory, according to an enthusiastic account of the Brusilov offensive written 15 years later by Russian general Nicholas N. Golovine (an anglicized version of Nikolai N. Golovin) in *The Russian Army in the World War*. None of the Allied powers could match the success of the Russian attack. The Austrian military nearly collapsed; it had to end its own offensive against Italy by transferring 15 divisions to the Eastern Front. Germany, fearful that the Austro-Hungarian Empire might sue for peace, sent 18 divisions from the Western Front and 4 reserve divisions that had been housed in Germany in order to bolster Austrian forces and keep them in the war.[25]

Airpower made an observable contribution to the success of the Brusilov offensive. The EVK used extensive photography to reveal fully the enemy's defensive order of battle. The Russian Southwestern Front employed 17 squadrons, comprising 90 pilots and 88 single-engine aircraft. (The last two numbers clearly illustrate the Russians' chronic problem of not having enough pilots and planes to operate the expected standard of 6 airplanes per squadron.) Nonetheless, Russian fighters hampered the ability of Austrian air reconnaissance to identify the point of Russian attacks; the fighters also tried to protect Russian aircraft that carried out intelligence-gathering missions. This combination of scouting and EVK photography enabled Russian artillery to suppress and destroy the opponent's defenses and to cause more damage with fewer cannon. During the breakthrough period of the Russian advance, the 17 air squadrons carried out 1,805 combat missions. Peak activity occurred in August 1916, when pilots completed 749 flights under battle conditions. By October the Brusilov offensive stalled, partly because of Russian casualties, autumn rains, and the fact that the Austrian line of defense had been greatly strengthened by the large number of new Austrian and German military divisions. Finally, the static nature of the Southwestern Front in the fall of 1916, coupled with the long-term stability of the Northwestern Front, explains the rapid expansion in the number of balloon detachments that year. By December there were some 73 balloon observer stations in 13 balloon divisions operating from the Baltic Sea in the north to the Black Sea in the south. Although balloons often were subjected to enemy gunfire, they played an important role in observing activity in the forward lines of German and Austrian troops.[26]

The achievement of Russian armies in advancing into Austrian Galicia was more than matched by the power and work of the Black Sea Fleet in checking the Central Powers and turning the sea into a Russian lake. First, the fleet continued to send hydrocruiser task forces and their shipborne flying boats to intervene and disrupt the transit of coal by attacking Turkish steamers and sailing ships. The effort proved so thorough that at times Turko-German ships, including the *Goeben* and *Breslau*, lacked the fuel necessary to steam into the Black Sea. By December 1916 the Russians had sunk or captured more than a thousand Turkish coastal craft. On one of the hydrocruiser visits to Zonguldak the following February 6 (N.S.) fourteen Grigorovich aircraft dropped thirty-eight bombs on the ex-German collier *Irmingard*—the largest vessel to be lost to an air attack in any theater of battle during the Great War. Second, task groups repeatedly bombarded the Bulgarian port of Varna. On August 25 (N.S.) the aircraft carriers *Almaz*, *Aleksandr*, and *Nikolai* sent nineteen planes into the harbor to bomb German submarines.[27]

Finally, in 1916 the army-dominated Stavka finally decided to take a step that it had refused two years before. Vice Admiral Andrei A. Eberhardt, an aggressive commander, had wanted from the beginning of the war to prepare the Black Sea Fleet for amphibious operations should the Ottoman

Empire become a member of the Central Powers. Even though Stavka had mistakenly predicted that the Turks would remain neutral, the army did not want to commit a substantial number of soldiers for waterborne military action. What we now call "if history" is fiction, of course, but if the Russian army had approved the preparation of amphibious troops with the Black Sea Fleet, such a force might have collaborated with the Allied attack at Gallipoli. British and French warships bombarded the peninsula in February 1915 and later landed troops there. A major amphibious assault by Russian soldiers disembarked from the Black Sea near Constantinople might have led to the occupation of the Turkish capital, knocked the Ottoman Empire out of the war, and brought Bulgaria, Greece, and Romania into the conflict on the side of the Allies. Black Sea ports would then have been opened to safe and copious trade and the entire chemistry of the war would have been altered.[28]

The Black Sea Fleet and its hydrocruiser task forces housing flying boats also assisted the Imperial Russian Army in its Caucasus campaign against the Turks. The navy and its Grigorovich aircraft interrupted, captured, or sank Turkish ships that carried troops and supplies eastward to the front against Russia. When an army-sized Turkish relief force under Vehip Pasha marched along the northern coast of Anatolia, Russian war vessels and aircraft harassed the troops and damaged supply columns, leaving the relief force no option but to retreat. Then, in March 1916, Stavka finally agreed to an amphibious operation against the Ottoman Empire. The dreadnought *Rostislav*, gunboat *Kubanets*, 4 torpedo boats, 2,100 soldiers (shipped on 2 transports), and 3 flat-bottomed minesweepers entered the small Atina harbor. Just behind Turkish lines, the amphibious exercise caught Russia's enemy in a surprising pincer that enabled the Russian Army to advance westward into the Turkish port of Rize on March 6 (N.S.).[29]

The Black Sea Fleet became heavily involved in augmenting Russian troops and supplies for the Caucasian Front. On April 7 (N.S.) approximately 16,000 Cossack soldiers were shipped to Rize on 36 smaller transports, with 8 flat-bottomed *Elpidifor* craft for the amphibious coastal landing stage. The substantial number of infantrymen had the protection of a dreadnought, 3 cruisers, and 15 torpedo boats. Three hydrocruisers holding 19 flying boats accompanied the naval task force. Aircraft provided a reconnaissance screen against German submarines and Turko-German warships. The additional troops enabled the Russians to stall and then defeat a serious Turkish counterattack. By April 19 (N.S.) the Russians had pushed the enemy westward and occupied the major Turkish port of Trabzon (anglicized as Trebizond). In the second half of May and early June, the Black Sea Fleet then used two convoys to transfer to Trabzon the 123rd and 127th infantry divisions, which included more than 34,000 men. Once again hydrocruisers used cranes to unload M-9s to the sea—planes that then flew scouting missions to help protect the convoys from enemy ships and U-boats.[30]

In July 1916 Vice Admiral Aleksandr V. Kolchak replaced Eberhardt as commander of the Black Sea Fleet. It would be nice to say that Eberhardt deservedly retired with honor, but the reality is that in both government and military, administrators and officers often engaged in politics and infighting to gain preferred appointments. Nevertheless, as a rear admiral Kolchak had been an excellent chief over the Baltic Sea's destroyer force, and he clearly valued aircraft now. While Eberhardt had established naval air stations at Batum, Rize, and Trabzon, Kolchak tripled the number of airplanes in some cases. On September 11 (N.S.) he dispatched flying boats to bomb the Bulgarian port of Varna as well as the Eukhinograd German submarine base. In August he also began secretly laying hundreds of mines around the Bosporus and later Varna; the minefields were constantly augmented, so that in essence the Central Powers were denied access to the Black Sea. It would be eleven months before the *Breslau* dared to steam through the Bosporus. Finally, Romania's entry into the war as a member of the Allies in 1916 led to that country's defeat by a German-Bulgarian-Turkish army under Mackensen. On December 16, 1916 (N.S.), Romania's main port of Constanța also was mined. The only Turkish vessels remaining in the Black Sea were smaller sailing ships berthed in lesser ports along the coast of Anatolia.[31]

CHAPTER 7

THE 1917 REVOLUTION IMPACTS SQUADRONS

★ ★ ★ ★ ★

D URING THE WINTER of 1917 the Russian Black Sea Fleet launched a series of important missions against Turkish naval forces. In early January the naval command sent 3 Russian dreadnoughts, 2 cruisers, and 3 destroyers to the Anatolian coast, with orders to take out Turkish sailing ships. They succeeded in sinking 39. Later that month the Russian cruiser *Pamiat' Merkuriia* and a couple of destroyers accompanied 2 large minelaying ships that placed an additional 440 mines near the entrance to the Bosporus. Between January 9 and March 7 (all dates from here on are N.S.) 5 Russian submarines—*Nerpa, Narval, Kashalot, Kit,* and *Morzh*—sank 26 more Turkish vessels. Later in March a task force with the hydrocruisers *Aleksandr* and *Nikolai,* and the *Romania,* a commandeered Romanian passenger ship, used their cranes to lower twelve Grigorovich flying boats into the sea near the coast just west of the Bosporus. Four of the planes bombed the Turkish waterworks at Lake Terko; the other eight conducted a photo-reconnaissance mission over the land northwest of Constantinople. The amphibious operations of 1916 had enjoyed such success that Stavka actually considered landing amphibious troops to capture the Ottoman Empire's capital city—a move that potentially could have forced the Turks to leave the war.[1]

Unlike the Black Sea, the Baltic Sea remained quiet during the winter months of 1917 because the gulfs of Riga and Finland had iced over. The armies on the Eastern Front remained less active as well. In good weather a reconnaissance plane on one side or the other would fly over enemy turf to see whether those forces might be preparing for a spring offensive. Nuisance weapon-firing also occurred, but muddy ground or swirling snow over the front discouraged any notion of trying to conduct a full-scale offensive. On the plus side, the Russian Army was fully armed; on the minus side, there were rumblings from mutinous soldiers who lacked boots and warm clothing. Regardless, Stavka had high expectations for the springtime advance. By the end of February, the tsarist empire had 545 combat-ready aircraft, augmented by 43 SPAD S.7s from France. France normally sent Russia out-of-date airplanes, but the S.7 was a potent fighter. It had a 200-hp Hispano-Suiza engine, a Vickers machine gun, and a pair of Le Prieur rockets. The first purchases went to Kozakov's First Combat Air Group on the Russian Southwestern Front.[2]

At the same time, the EVK, now headquartered in Vinnitsa, sent a detachment 210 miles almost directly south, near the Romanian Front and the town of Bolgrad, with plans to send large bomber-reconnaissance aircraft to assist in a spring push by the Russian Army to liberate both Romania and Serbia from Austrian and German troops. But all of Stavka's big plans fizzled. Two developments quelled those ambitions. First, over the winter Russian soldiers in dozens of units up and down the Eastern Front began to make it clear that although they would fight to defend existing front lines they would not participate in any new offensives. It was not just a matter of lacking boots and warm clothes during winter. Too many comrades had died for no apparent reason. Even Brusilov's offensive had faltered; repeated Russian assaults in the marshes outside the town of Kowell had led to massive deaths against an entrenched enemy and German firepower.[3]

Adding to the soldiers' sorrows and fears of injury and death associated with combat was the news from home that was delivered by replacement troops and through the occasional letter from family members or friends. Beyond the ugly lies and actual truths concerning the tsar and the government were the disquieting realities about war weariness among the Russian populace. Aside from the horrendous and disruptive loss of millions of soldiers that affected families on the homefront, civilians struggled with rampant inflation. The war forced the tsarist government to end gold as the foundation of Russian currency. By 1917 a loaf of bread cost a ruble—a sharp rise from just a few months before, when it could be bought for kopecks (pennies). Workers' wages remained the same despite the large increase in their living costs—a factor that led to numerous strikes and work stoppages. Money was not the most serious issue when it came to food for city dwellers, however. The economic exchange system between town and country had collapsed. Large landed estates lost hired laborers to the army. Peasants farming the land—now only old men, women, and children—no longer could trade grain, vegetables, and meats for clothes and agricultural implements. As a result, peasants grew less and kept more to feed themselves and their animals, if they had any.[4]

On March 8, 1917, bands of housewives, war wives, and women workers greeted International Women's Day to protest the bread shortage in Petrograd. Striking male workers locked out of factories soon joined their wives, sisters, and mothers in clashes with police. Within a few days the movement grew and evolved into uncontrolled revolution. Police disappeared, Cossack horsemen refused to charge into crowds, and troops assigned to bring order out of chaos sided with the women. After Tsar Nikolai II dismissed Russia's national legislature—before he abdicated the throne—some Duma members who remained in the assembly's Tauride Palace formed a committee that led to a new entity called the Provisional Government. There were two obvious problems with this replacement administration. First, no one, even members of the Duma, had voted the Provisional Government into office; whatever support it had came from the minority of bourgeois citizens. Second, ministry heads clearly understood that their governing role would be temporary, and so avoided addressing any of Russia's terrific problems as the country awaited the meeting of the Constituent Assembly. For a variety of reasons, the assembly did not convene until January 1918; the Bolsheviks dissolved it after one session.[5]

Meanwhile, four days after the initial uprising, when the committee that soon created a Provisional Government was formed in the Tauride Palace, a second shadow government emerged in another wing of the same building. It patterned itself after the Council (to be called a Soviet from then on) of Workers' Deputies that had appeared during the 1905 Revolution. Among other factors, it was part of the background that had prompted the tsar to respond politically to the 1905 Revolution by creating a Duma and revising the empire's Fundamental Laws of 1832. In 1917 the Petrograd workers' Soviet was quickly inundated by soldiers selected locally by military units in the city. Initially, radical Marxists (Bolsheviks who later adopted the name Communists) were a tiny minority. Regardless, the Petrograd Soviet had three missions—to keep a critical eye on the Provisional Government, to defend lower-class interests, and to prevent military officers from using the approximately 250,000 troops then in the capital to crush the revolution. Thus, the Soviet issued Order No. 1, which required troops in each military group of the armed forces to elect a committee. The committees or soviets were to ensure that all political matters of their units were subordinate to the Petrograd Soviet, not to army officers.[6]

The Provisional Government not only approved Order No. 1 but agreed to its empirewide distribution through the Petrograd Telegraph Agency—the precursor of TASS (Telegraph Agency of the Soviet Union). As soldiers and sailors learned of Order No. 1 and formed committees or soviets they ignored that part of the document upholding military discipline. Instead, the order was interpreted as bringing democracy to the armed forces, which ended military obedience. Many soldiers and sailors on the front lines could support Russia defensively, but they absolutely rejected any action related to fighting, dying, or suffering casualties by attacking the enemy. By contrast, most aviation squadrons in the army or navy continued to gather intelligence, fight opposing aircraft, or bomb targets of opportunity. Several pilots actually reached ace status after the Russian Revolution. It must be remembered, however, that pilots generally were officers, not conscripted soldiers; many of them held the formal designation of Military or Naval Pilot, which required them to take an oath of service. Unlike the conflict involving ground forces or naval ships in the Gulf of Finland, the air war continued.[7]

This is not to suggest that pilots and squadrons were unaffected by revolutionary events. Massive desertions occurred later. Reserve troops serving away from the front lines began to take advantage of the evaporation of army discipline by drifting away from their units. Most were peasants, who preferred to be home in time for spring planting rather than risk becoming a casualty of war. Conscripted soldiers at air squadrons, who had provided muscle, maintenance, and security for operating aircraft, also began leaving their posts. (Air squadrons were usually behind the reserve troops and even farther away from battle trenches and artillery.) That meant that pilots had to load bombs, fill cartridge belts, pour fuel, push planes into or out of storage positions, and swing propellers to start engines. Sergeants who served as mechanics in maintaining airframes and engines had to be bribed—in some cases with better food, housing, and pay—to provide those services. The rampant inflation exacerbated the problem. It took seven rubles in 1917 to purchase what one ruble bought in 1914. As the ruble declined and the economy collapsed, the arrival of fuel, food, armaments, and aircraft replacement parts became erratic; military pilots and their mechanics had to find alternative sources for supplies that grew scarce as the war dragged on.[8]

Disappearance of military discipline also occurred everywhere, as Aleksandr Riaboff discovered. The pilot completed his flight training at Gatchina, received a leave of absence to visit his family, and then took a train to Odessa several months after the Russian Revolution; he was genuinely surprised that the railroad still functioned. Riaboff had been assigned to advanced pilot training in Nieuport 17 fighter aircraft at the School of High Pilotage, located twenty miles southwest of Odessa near the village of Lustdorf, which had been settled by Germans during the reign of Catherine the Great (1762–1796). The commanding officer, Staff-Captain Aleksandr A. Chekhutov, treated the fifty student pilots and several instructors and mechanics like one big happy family on a long vacation that, for fun, included flying airplanes. Flight times were indefinite and instructors were neither strict nor punctual. To Riaboff, their behavior perfectly matched "the general chaos and anarchy that prevailed" in revolutionary Russia. The student pilot lived a life of bliss. He and Valerian Przhegodsky, his barracks roommate from Russian Poland, often went to the shore of the nearby Black Sea for a swim.[9]

Although the revolution made Riaboff's flight training at the School of High Pilotage blissful, the upheaval also resulted in less-pleasant events. In the Gulf of Finland, for example, sailors often gave rough, sometimes deadly treatment to their officers. Most sailors were conscripted from factories and brought Marxist ideas with them to the cruisers and dreadnoughts. To protect the gulf and capital city from German warships, Russia's large vessels generally remained idle during the war. The sailors proved to be one of the most radical groups in the former empire. They protected the Bolshevik Party of Lenin during the November Revolution and for most of the Russian Civil War. The fact that ships in the Gulf of Finland were inactive for more than two years did not prevent officers from using a heavy hand in disciplining sailors and denying them shore leave, and the sailors often retaliated harshly. Once, revolution-liberated sailors got even for their perceived abuse by murdering about seventy-five high-ranking officers. One of the early victims was the Baltic Sea commander, Vice Admiral Adrian I. Nepenin. The sailors also formed unions and soviets, abolished titles of warrant officers, removed shoulder boards from officers' uniforms, created a sailor-driven Central Committee of the Baltic Fleet, and held a Baltic Fleet Congress in Helsingfors (today, Helsinki, Finland).[10]

Although the formal relationship between naval officers and sailors changed dramatically in both the Gulf of Riga and the Black Sea, cooperation and service continued for a time because of the mutual respect that the two sides had developed during successful and sometimes dangerous wartime missions. For air combat around the Gulf of Riga, the navy began to train pilots in land-based pursuit aircraft. In May of 1917, for example, Alexander Seversky was ordered to attend the Moscow Aviation School at Khodensk Aerodrome. Several navy pilots began flight training in Nieuport 17 fighters. Circumstances and technologies caused a shift away from using flying boats exclusively. Not only had German ground forces completely occupied the gulf's western shore, but they built a large naval air station at Windau, within flight range of the Russian naval air stations on Ösel Island. The Germans at Windau could house and launch advanced seaplanes such as the Albatros W.4, which entered production in the summer of 1916. The new Albatros had more speed and greater maneuverability and carried two forward-firing machine guns. The M-11 developed by Dimitrii P. Grigorovich, also in 1916, was intended to be a flying-boat fighter aircraft, but unfortunately it was underpowered and too slow. It was another Russian shortcoming in aircraft motors.[11]

At the beginning of April 1917 the navy signed a contract with Russia's largest aircraft manufacturer, Moscow's Dukh Company, to build seventy-five Nieuports. The contract specified fifteen dual-controlled trainers, fifteen two-seat reconnaissance aircraft, and forty-five single-seat fighters—Model 17 and the 21 variant. The latter was slower, but its larger control surfaces made it more maneuverable, a key benefit in a dogfight. Regardless, both aircraft flew faster than the German seaplanes that Russian pilots would face over the Gulf of Riga. In June, when Seversky graduated from his course at the Moscow Aviation School, the naval air station at Zerel on Ösel Island experienced an attack that fulfilled the navy's worst fears about Russia's loss of airpower superiority in the Gulf of Riga. At Zerel, the Germans bombed the air station's sheds, destroyed two flying boats, and damaged four others. Russian aircraft that survived the German bombs were too slow to defend the air station and could not pursue the enemy aircraft once the German bombing runs had ended.[12]

After Vice Admiral Nepenin lost his life from a sailor's bullet, Vice Admiral Aleksandr Kolchak survived to command a strong Black Sea Fleet. In one sense, his fleet resembled the activity of ships and aircraft in the Gulf of Riga, on a much larger scale. Indeed, Kolchak had an abundance of ships and planes. The revolution found him with seven hydrocruisers, stocked with thirty-eight flying boats, with ninety-four other military aircraft in reserve at four land stations. Another twenty aircraft functioned as trainers for future pilots. In April and May hydrocruiser task forces used flying boats to bomb Turkish batteries or provide scouting cover against Turko-German warships as Russians set new mines around the Bosporus entrance. As happened in the case of army air squadrons, naval supplies became scarce over time, which caused delays in repairing or resupplying aircraft. Worse yet, cities like Odessa that were near Black Sea shores formed soviets that matched or exceeded the radical nature of those in northern cities. On June 19 Kolchak was removed from his command (though without being shot) by a sailors' soviet that also disarmed all their officers. Four days later the Turko-German light cruiser *Breslau* steamed into the Black Sea for the first time in 1917. It served as a symbol of the revolution's impact on Russia at war.[13]

The revolution's effect on the Squadron of Flying Ships was palpable when a weak Provisional Government and a strong soviet movement across the former empire replaced the tsarist regime. At Vinnitsa the EVK created workshops and filled supplies into a factory and warehouses that had once belonged to a German company. On March 15, when the tsar abdicated, an arsonist who opposed the planned spring offensives against German and Austrian troops on the Southwestern Front started a fire in one of the warehouses. Within a month, all the soldiers who had provided muscle and security for the large aircraft had left the aerodrome—influenced by a speech on land and freedom delivered by a member of the Petrograd Soviet. Shortly afterward, the Provisional Government forced Mikhail Shidlovskii to resign as leader of the EVK. In protest, Igor Sikorsky also departed. Although the Il'ia Muromets planes continued to fly, tragedy struck in May. Georgii I. Lavrov, who replaced Shidlovskii, fell victim to a saboteur who weakened a strut and its arresting cables on the four-engine plane that Lavrov was piloting. He and his crew died in a fatal crash near the town of Mikulinsta. A technical committee associated with the Directorate of the Military Aerial Fleet erroneously concluded that the plane was unsafe. Nevertheless, Il'ia Muromets aircraft already assigned to the Eastern and Romanian fronts continued their operations.[14]

The soviets had become such a broadly based feature of the revolution that a national Congress of Soviets met in Petrograd in June. At that point only 15 percent of the elected delegates belonged to the Bolshevik Party. Moderate socialists dominated the Central Executive Committee, which actually confirmed soviet support for the Provisional Government. In fact, however, Russian society and the economy had descended into an unsettled and radical decline. During the congress Petrograd workers, soldiers, and sailors conducted massive demonstrations against both the war and the Provisional Government; in urban settings workers conducted strikes or took over factories and often removed or supervised managers; in rural areas peasants started to seize land, animals, and farm tools from gentry estates in what amounted to a remarkable confiscation of property. Clearly, Russia, with its dual authority, neglected to address its most serious problems. And the Constituent Assembly, which was expected to create a decent constitution and a genuine government, failed to appear because a minority of voters had fled from—or lived under occupation by—the enemy.[15]

Meantime, almost six months before the Constituent Assembly opened there were huge demonstrations, especially by soldiers and sailors against war and the Provisional Government. Aleksandr F. Kerenskii, soon to be the head of the Provisional Government, held the ministry posts for both the army and navy. As such, he came under heavy pressure from Allied nations for Russia to resume fighting on the Eastern Front. The virtual end to combat potentially opened the possibilities that Austria might transfer troops to battle Italy and that the Germans would bolster their armies in the Western with soldiers from the Eastern Front. In May Kerenskii sent out an order calling on forces to resume military discipline and prepare for combat. He followed that up with tours of the front lines, where he delivered speeches to inspire troops to defend the revolution through offensive action that would remove the enemy from Russian territories. Adversaries described him unkindly as the "persuader in chief," but he was a spellbinding orator. Nevertheless, not long after Kerenskii departed it became obvious to officers that their men had little interest in fighting and dying in planned offensives.[16]

Unlike ground forces, pilots and planes were fully prepared to participate in offensives from the Northwestern to the Romanian fronts. Since the Southwestern Front had enjoyed great military success under Brusilov in 1916, it served as the showcase for the assault to protect the revolution and liberate Russian lands from enemy occupation. Six Il'ia Muromets planes in 2 detachments, the First and Third, flew into the rear echelon of the Austrian-German military and bombed reserve troops and railroad facilities. Furthermore, at least 225 single-engine airplanes were focused on 2 sectors around Lutsk and Kosovo. Along the front, 1 aircraft was assigned for every 0.5 kilometer (0.3 mile). To lay the foundation for the offensive, numerous photographs were taken of enemy troops and artillery batteries. Processed images at 2 laboratories supplied information that was transferred to maps and distributed to officers who headed the infantry attack. Once the heavy artillery fire began, scouting aircraft corrected and directed the aim of Russian gunnery. It should come as no surprise that half of Russia's ace pilots flew with fighter squadrons on the Southwestern Front, including Pavel V. Argeev, Georgii U. V. Gil'sher, Aleksandr A. Kozakov, Evgraf N. Kruten', Donat A. Makienok, Ivan A. Orlov, and Vasili I. Ianchenko.[17]

The middle, or what the Russians called the Western Front, played a secondary role. Responsibility to press against the enemy and move the front farther to the west went to the Russian Tenth Army. Nine aircraft squadrons assisted forces on the ground, which consisted of four corps squadrons, one army, one artillery unit, and three fighter detachments of the Third Battle Group. Altogether fifty-nine pilots and thirty-six observers had access to sixty-one aircraft, which meant that each squadron initially had its full complement of six planes. The squadrons helped ground forces prepare for the offensive by photographing the first and second German trenches multiple times. Strategic German areas in the rear, however, only received two visits from Russian reconnaissance—a lapse that resulted in uncertain intelligence about the latest placement of reserve troops and military supplies. Even so the preparation period, which involved 535 flight missions, proved costly. Despite the best efforts of the Third Battle Group, more than a dozen pilots and observers died, and others suffered wounds from the machine guns of German planes and from antiaircraft fire on the ground.[18]

The Romanian Front served as a mere auxiliary to the Southeastern Front, where Russian troops began their artillery barrage on July 1, 1917. Yet the Russian attack on the Romanian Front started later and lasted longer. It also contained the second largest number of army ace pilots—Ivan A. Loiko, Aleksandr M. Pishvanov, Ivan V. Smirnov, Vladimir I. Strzhizhevskii, and Grigorii E. Suk—who were responsible for several of the twenty Austrian planes that were shot down during the battle. Aviation detachments came mainly from the Russian Fourth and Sixth armies. Before the offensive began four squadrons spent weeks gathering intelligence and taking photographs. Most of the reconnaissance missions were conducted by the Twenty-Seventh Air Corps, the Fourth Army Air Artillery, and the Free-Romanian Air Squadron. The Fourth Army Fighter Air Detachment protected scouting planes from enemy aircraft attacks. The Russian Sixth Army deployed experienced observers in two airplanes equipped with radio communications that enabled them to provide immediate firing corrections to the Russian batteries. When the Russian Fourth Army initiated the offensive, it acquired another squadron and a fighter detachment to use for scouting and protection.[19]

Finally, the Northwestern Front had the task of mounting a diversionary attack against the Germans. The commander, General Vladislav N. K. Klembovskii, possessed a decent and well-equipped air force for his Russian Twelfth Army. It contained three corps squadrons (the Tenth, Twenty-Third, and Thirty-Third), one army squadron, and two fighter detachments. In a small way, however, the experience of Staff-Captain Konstantin K. Vakulovskii was emblematic of what happened to Kerenskii's offensives along the Eastern and Romanian fronts. In charge of the First Fighter Detachment, Vakulovskii flew over enemy territory on July 1, 1917. As he photographed the third line of the enemy's defense, the Germans greeted him with antiaircraft artillery. A shell that exploded below his Morane-Saulnier airplane knocked out his engine. The damaged motor leaked gasoline as Vakulovskii piloted his powerless glider eastward to the Russian position. By the time he landed, a fire had broken out in the cockpit, burning his jacket and his right arm. Happily, the pilot survived, recovered from his wound, and would fly again as an ace pilot, but the Russian Twelfth Army's diversionary attack simply failed.[20]

Indeed, the Russian attacks against Austrian and German troops failed everywhere. The most obvious and painful of these shortcomings came on the Southwestern Front. Many Russian soldiers refused to advance, and reserve units refused to participate. Soldiers who did advance walked casually, took some souvenirs, and then returned to their own trenches. Many held meetings in which they decided not to take part in the offensive and not to obey military orders. In some cases groups of soldiers decided instead to assault or murder their officers. Some infantry troops finally did move forward; Russian artillery there laid down such a barrage of ordnance that opponents died, suffered injury, or retreated. Later in July the Germans took advantage of their enemy's weakness; they conducted a counteroffensive directed at the Russian Eleventh Army. Russian soldiers fled helter-skelter from their trenches. Remnants of the Russian Army continued to exist, but would no longer fight the enemy. Kerenskii's notion that his front-line oratory had created a new and powerful revolutionary army

Tsar Nikolai II speaks with designer Igor Sikorsky, who stands in the cockpit doorway of the *Grand* aircraft in 1913. *Courtesy of the National Air and Space Museum, Smithsonian Institution (NASM 90-2091)*

Petr N. Nesterov stands next to a French-built Nieuport IV monoplane in 1913. *Courtesy of the National Air and Space Museum, Smithsonian Institution (NASM 90-8302)*

Informal portrait of civilians inspecting a seaplane, circa 1913. *Courtesy Von Hardesty*

Alexander P. de Seversky (Aleksandr Nikolaevich Prokof'ev-Severskii) after his solo military flight in a Russian-built Farman in 1915. *Courtesy of the Cradle of Aviation Museum*

M-5 seaplane with Vickers machine gun. Black Sea Fleet, circa 1915. *Courtesy Von Hardesty*

FBA naval flying boat built under Franco-British Aviation license in Russia in 1915. *Courtesy of Aeronaut Books*

Il'ia Muromets reconnaissance-bomber at Iablonna, with EVK squadron members on wings in 1915. *Courtesy of the National Air and Space Museum, Smithsonian Institution (NASM 90-2144)*

Grand Duke Aleksandr Mikhailovich (*second from right*) visits Major General Mikhail V. Shidlovskii (*third from right*) at the EVK squadron at Iablonna in 1915. *Courtesy of the National Air and Space Museum, Smithsonian Institution (90-2167)*

An FBA naval flying boat is lifted from the deck of a Russian ship in 1915. *Courtesy of Aeronaut Books*

Aleksandr A. Kozakov poses with his Russian-built Nieuport 10 in 1916. The ace of Russian ace pilots in the Great War, he eventually flew with the British in the Russian Civil War. *Courtesy of Aeronaut Books*

Dimitrii P. Grigorovich designed Russia's best naval flying boat, the M-9, which went into production in 1916. The M-9 continued to be used during the early Soviet period. *Courtesy of Aeronaut Books*

Viacheslav M. Tkachev is shown with a Russian-built Morane-Saulnier aircraft and its parasol wing in 1916. During the Russian Civil War he served as head of aviation for anti-Soviet Russians in the Crimea. *Courtesy of Aeronaut Books*

The last or E series of the Il'ia Muromets emerged at the end of 1916. Several of these aircraft became part of the air arm of Soviet Russia. *Courtesy of Aeronaut Books*

Andrei N. Tupolev designed Russia's first all-metal bomber, the ANT-4, TB-1, in 1924. *Courtesy of the National Air and Space Museum, Smithsonian Institution (NASM A-47057-D)*

completely evaporated. Even the idea of preserving a strong defensive line had vanished. Especially among soldiers, sailors, and workers, the Provisional Government lost all respect. Meanwhile, the Bolshevik Party rapidly gained new members as it called for an end to the war and "All Power to the Soviets."[21]

The relatively quick end to the Russian attack—and especially the German counteroffensive on the Southwestern Front—had a significant effect on the EVK. When Russian troops fled Tarnopol, not far from EVK headquarters at Vinnitsa, Russia needed to make plans to move the base eastward, out of harm's way, but the rapidly developing anarchy blocked any such effort. Airplanes could fly to a new headquarters, but trains needed to carry supplies, engines, airframes, fuel tanks as well as repair kits, materials, tools, and machines, but railroad transportation had become unreliable. Fortunately, the disappearance of Russian soldiers inadvertently saved Vinnitsa; the German counteroffensive temporarily stopped before it reached the EVK headquarters. Regardless, for a brief time, twenty-two Il'ia Muromets aircraft remained active up and down the front as four primarily reconnaissance detachments kept track of German activity (or inactivity). Except for attacking the Northwestern Front, the Germans chose not to make a broad-based military incursion into Russia. An offensive would be silly. Logically, the Germans waited until it was clear that some type of official governing body existed in Russia that could sign an armistice and negotiate a treaty that would end the Great War in the east. That moment came on December 15, when the new Soviet government agreed to a cease-fire with the Central Powers.[22]

That same December eight flying boats in various stages of disrepair assembled at Vinnitsa. When the Russians walked out of the negotiations with the Central Powers on February 10, 1918, the Germans were shocked. They soon recovered, though, and on February 18 they began an unopposed offensive. Among other places, the Germans occupied Vinnitsa and the aerodrome that housed the EVK's headquarters. Remaining Russian crew members set fire to the Il'ia Muromets planes so that the Germans could not take them as prizes of war and add them to their air force inventory. Four of the reconnaissance-bombers ended up in the hands of Polish and Ukrainian military units. Ten others stayed with the dwindling Russian

Army. Because of the fallout from the revolution, production of Il'ia Muromets aircraft ceased, and Vinnitsa had no repair supplies. That meant that several flying ships in the care of the army had to be cannibalized for parts and engines in order to keep several versions available for flight. Lieutenant Colonel Aleksei V. Pankrat'ev, the former assistant EVK chief, volunteered to work for the new Soviet government. He joined the Worker and Peasant Red Military Air Fleet and subsequently became the commander of the Flying Ship Division.[23]

Back in June 1917, Seversky did not return to his squadron after graduating from the Moscow Aviation School. Instead, he remained at the school as a flight instructor who trained navy pilots in Nieuport 17 aircraft. In July, after the failure of Russian offensives and the success of Petrograd demonstrations that favored Soviet governance, Seversky received a telegram from Captain (soon to be Rear Admiral) Boris Dudorov. The captain had served as chief of the Naval Aviation Department of the Naval General Staff in Petrograd. Lieutenant Seversky was ordered to leave Moscow immediately by train and report directly to Admiralty headquarters, located next to the Winter Palace, then home to the faltering Provisional Government. Dudorov had a problem. The Dukh Company in Moscow had completed major parts for the Nieuport 17 and 21 and transported them to Petrograd, where they were put together by the Shchetinin Aircraft Company to create a finished plane that could fly to naval air stations on or near the Gulf of Riga. Like most factories, however, Shchetinin suffered work stoppages carried out by workers who demanded more pay, better hours, and in some cases worker supervision of management.[24]

When Seversky arrived at the Admiralty, he was assigned by Dudorov and the captain's aide, Aleksandr A. Tuchkov, to go to the Shchetinin factory and convince workers to return to their stations and complete the navy's desperately needed aircraft. There were two reasons that Dudorov had picked Seversky for this task. First, the lieutenant was a well-known hero, who had shot down German planes in the Shchetinin-built M-9 flying boat. Second, earlier, when Seversky was getting used to wearing a wooden prosthesis on his right leg, he had been appointed chief naval aircraft inspector for Petrograd. At Shchetinin he had spent time with workers and with Dimitrii Grigorovich and had argued that the newly designed M-9 should possess a rotating machine gun and some armor to protect the crew. In his July 1917 pep talk to the workers, he pointed out that the navy and its air arm had not left their position in the Gulf of Riga. They helped prevent Petrograd from being attacked from Riga by the Germans. "[W]hen the new government [by the Constituent Assembly] is organized which will be representative of your ideas, that government can decide the question of peace or war," Seversky went on. "Until then, let every one of us ... do our duty and support each other to the fullest extent."[25]

Seversky's talk proved effective. Workers cheered and carried him to the waiting car that would take him back to the Admiralty. Captain Dudorov acknowledged the speaker's success by reappointing Seversky to his previous command of the Second Pursuit Squadron at Zerel. By the end of July the detachment had secured four Nieuport 17s, one Nieuport 21, one Grigorovich M-9, and six Grigorovich M-15 flying boats. Additional Nieuports soon became available to help several squadrons on Ösel Island in their role on the center stage of the last Russian battle against the Germans during the Great War. Seversky's squadron had the task of protecting the battery of four 12-inch guns at Zerel that stopped German surface vessels from steaming into the Gulf of Riga through the Irben Strait. Key support focused on preventing German minesweepers from removing Russian mines that were keeping German battle cruisers from entering the straits and bombarding the Russian battery. The Nieuports protected the M-15s that had dropped bombs to harass the enemy's minesweepers—a precursor of the use of fighter aircraft in World War II. Naturally, German seaplane patrols tried to screen the minesweepers, resulting in numerous air battles in August.[26]

By the end of August, the First Pursuit Squadron at the Arensburg naval air station also was equipped with Nieuport 17 fighter aircraft. Located northeast of the Sworbe Peninsula and Zerel station, the Arensburg detachment was commanded by Lieutenant Mikhail I. Safonov. His squadron hunted German submarines and attacked reconnaissance planes that tried to gather intelligence on Russian warships. German seaplanes also attempted to drop bombs to disrupt the air station's work, but Russian Nieuports and M-15s warded off or shot down enemy aircraft. By the fall Safonov would join Seversky as an ace pilot. The success of the two squadrons would be diminished by the triumph of German land forces. The German Eighth Army advanced on Riga against a dwindling and demoralized Russian Twelfth Army. Token Russian resistance evaporated as soldiers fled from the field of battle, leaving their guns behind them. On September 3, German troops occupied Riga, along with the former Russian fortress of Dünamünde.[27]

The German victory in taking Riga had multiple repercussions. First, the fleeing soldiers also forced the substantial air arm of the Russian Twelfth Army to make a hasty retreat. The air arm, typical of Russian forces in 1917, consisted of three corps squadrons (Tenth, Twenty-Third, and Thirty-Third), one army squadron, and two fighter detachments. The abrupt and unplanned evacuation led not only to the destruction of aviation property at the aerodrome but to the escape of aircraft to unfamiliar or natural fields, which resulted in a large number of aircraft suffering damage and occasional destruction. Second, the Germans converted Riga into a submarine port. Boats such as the *UC-57* and *UC-78* turned the tables on the Russians by laying mines that interrupted the Russian warships and supply vessels that operated around Ösel, Moon, and Dägo Islands at the entrance to the Gulf of Riga. Finally, on September 18, 1917, German Headquarters approved an attack known as Operation Albion, designed to rectify the Irben Strait problem by crushing Russian air, sea, and land forces in the area. The goal was to open Riga as a supply port for German transports as the army prepared to end the war in the east by sending troops to occupy Russia's capital city of Petrograd.[28]

Meanwhile, German seaplanes regularly visited Ösel Island in September to do as much damage as possible, not only to the First and Second air squadrons, but also to the Third, which was located at Kilkond. On one raid a German plane enjoyed a lucky moment. The pilot avoided getting shot down and dropped a bomb that hit the munitions depot at Zerel. Loaded shells exploded, killing 40 garrison troops and wounding many others. Fortunately for the Russians, the battery of guns survived intact and the shells could be replaced. On October 10, however, a German armada of 300 ships, commanded by Rear Admiral Ehrhardt Schmidt, approached the northwestern shore of Ösel Island. The German fleet entered Tagga Bay behind a screen of minesweepers. Nineteen of the transport vessels held

24,000 troops of the Forty-Second Infantry Division and the Second Bicycle Brigade. The massive flotilla was screened by 6 dirigibles and 109 aircraft. Confronting this huge force, the Russians had available only 30 aircraft, 12,000 unreliable soldiers, 2 dreadnoughts, 3 cruisers, 8 destroyers, and a small collection of gunboats, minesweepers, minelayers, and supply ships. Russian sailor committees on a number of other destroyers voted not to enter a naval battle with the Germans.[29]

Russian forces came under the command of Rear Admiral Mikhail K. Bakhirev, who proved to be, under such dire circumstances, an excellent leader. One major advantage that Bakhirev held was his role as the officer in charge of mine defense: he knew exactly where his ships could maneuver around the islands and through the entrance to the Gulf of Riga. Even so, at Tagga Bay German warships bombarded and silenced the two batteries; on October 12 German motor launches landed troops on shore, as most Russian infantry troops surrendered or fled rather than engage the Germans in a firefight. With enemy troops on the ground and virtually intact, Bakhirev understood that all three naval air stations on the island would soon face German forces. He ordered Russian pilots to fly their planes to the naval air station at Kuivast on the east side of Moon Island. The commanders of the first and second squadrons volunteered to stay at Zerel. The two friends, Safonov and Seversky, could conduct reconnaissance flights to keep Bakhirev informed about German activity in the Irben Strait. Almost surprisingly, the Russian garrison remained by the guns. It meant that food, fuel, ordnance, and even a wireless were available to the island's only remaining airmen.[30]

Courageously, Safonov and Seversky flew a number of reconnaissance missions into the Irben Strait on October 12 and 13. Several times the pair fought brief air battles in their Nieuport 21s against small enemy air patrols. They feared that without the rest of their squadrons more German seaplanes might arrive and end their scouting activities, so they planned to break off their engagement and use the speed of their planes to return unscathed to their air station. After hours of flight and combat, their faces were blackened by the castor oil from the Nieuports' Gnome rotary engines. The news that they had to forward on the wireless to Bakhirev was not good. They noted that a fleet of German minesweepers had gathered along the Courland coast, out of range of the Zerel battery. There were no M-9s or M-15s that could drop bombs and damage or force the minesweepers to retreat. During the night of October 13–14, in darkness and rainy weather, German minesweepers cleared a water path that enabled the dreadnoughts *Friederich der Grosse*, *Kaiserin*, and *König Albert* to approach the tip of the Sworbe Peninsula. As the ships fired their guns at the Russian battery, most of Ösel Island had been occupied and German soldiers were about to enter the peninsula. Safonov and Seversky had no choice but to fly their Nieuports out of Zerel to avoid the destruction of the planes by the bombardment or capture by the Germans.[31]

Safonov flew to Kuivast without incident, but Seversky's engine failed shortly after takeoff. Fortunately, he crash-landed away from enemy troops and near an Estonian village. The peasants hated the Russians, but they hated Germans even more. For two nights Seversky, with only one good leg, hobbled slowly across Ösel Island and avoided German soldiers, thanks to his Estonian protectors, who knew the land and how to keep the pilot hidden. At water's edge, Seversky liberated a rowboat and, despite being fired at by German soldiers, crossed the channel to Moon Island and safety. During the interim, the German navy entered the Irben Strait and, on the morning of October 17, attacked the Russian fleet. The gun battle damaged the dreadnought *Grazhdanin* (Citizen); Admiral Bakhirev's battle cruiser, the *Baian* (Accordion); and the cruiser *Slava* (Glory), which had to be scuttled. German naval power forced the admiral to order the withdrawal of Russian personnel from Moon, Dägo, and Worm Islands. His fleet then sailed north to the Gulf of Finland, leaving a trail of mines behind the ships. Despite the loss of the Gulf of Riga, Safonov and Seversky were treated like heroes. The navy promoted Safonov to senior lieutenant and placed him in command of the rejuvenated Second Pursuit Squadron. He subsequently shot down two German aircraft over Moon Island in November. Safonov's victories in the air may represent Russia's last aviation triumph in the Great War. Promoted to lieutenant commander, Seversky took charge of the several pursuit squadrons near the Baltic Sea.[32]

CHAPTER 8

REDS VERSUS WHITES

★ ★ ★ ★ ★

SAFONOV AND SEVERSKY were emblematic of a category of pilots who survived the war but chose to leave the portion of Russia that was controlled by the Soviet government. As members of the nobility and Imperial Russian Naval Academy graduates, they were lucky to have avoided being punished, pummeled, or killed in a revolutionary society that was determined to eliminate the upper classes. The two men survived because they were cautious about where they went, and it helped that both were noted for their heroic activities during the war. Like publications printed in the other belligerent countries, Russian newspapers and magazines gave favorable, sometimes adoring attention to pilots and their military service to the country. Both men separately remained at their posts after Russia signed an armistice with the Central Powers. That document was provisional in stopping the active fighting, and it did not end the war. On February 11, 1918, the Soviet government chose not to wait for a final peace agreement. It went ahead and demobilized the army and navy. True, the Council of People's Commissars did lay the foundation for the eventual formation of the Red Army and Red Fleet. Unfortunately, however, neither Safonov nor Seversky could be certain what role, if any, he would play in the new Soviet military. Safonov, for one, was actually discharged.[1]

Near the end of 1917 Finland, a portion of the former Russian Empire, had become an independent country, and Safonov decided to take advantage of that. It was very unusual for the Soviet government to surrender Russian-held land willingly. Even now, in the twenty-first century, many Russians want to reabsorb or at least control territories that were lost by the Russian or Soviet empires. Before the Great War, Finland had been a semi-autonomous grand duchy, with a Parliament (Diet) and a Russian population of less than 1 percent. There had been efforts to Russianize Finland, but such measures generally had failed. Regardless, Finland's grand duke was Nikolai II, the Russian tsar who no longer ruled the Russian Empire that previously had included Finland. As a result, in January 1918 the Soviet government temporarily recognized Finnish independence. The new country had the unfortunate experience of having endured a civil war between socialists and conservatives. During 1918 the conservatives, or White Guard, won under the leadership of General Carl Gustaf Emil von Mannerheim. To help gain victory, Finnish agents recruited several Russian pilots; among them was the very available Safonov.[2]

Safonov felt no guilt whatsoever when he absconded with an unused, two-place army Nieuport. In April he flew to Finland with his bride, Ludmila, who was buckled in the observer's seat. His flight took place at the same time that Seversky and his mother, Vera, were on a passenger vessel steaming from Tokyo to San Francisco. Back in October 1917, the heroic Seversky had received not only a promotion by the navy but also the honorary post of governor of the All-Russian Aero Club. Moreover, Aleksandr Tuchkov, the assistant head of navy aviation, had asked Seversky to join him in traveling to the United States as a member of the official naval mission attached to the Russian Embassy in Washington. But higher officials in the Admiralty prevailed upon the young pilot to delay his departure indefinitely; Petrograd was in immediate danger of being attacked by the enemy. The December 15 armistice changed nothing. Only a peace treaty signed by Russia, Germany, and the Central Powers could guarantee that the Russian capital would be spared from occupation.[3]

Life in Soviet Russia had become really complicated for Seversky when the navy was abolished, shortly after Commissar of Foreign Affairs Lev D. Trotskii (known popularly in the West by his anglicized name, Leon

Trotsky) had walked out of the peace talks. And Seversky was not alone in facing a dilemma; the Soviets suddenly had to transfer their capital from Petrograd to Moscow to avoid the unopposed German offensive. For obvious reasons, Lenin advocated signing the draconian peace treaty with Germany and the Central Powers. Other voices in the Soviet government argued for moving the capital east of the Ural Mountains and into Siberia and rebuilding the army. In this threatening moment when a one-sided war resumed, Lenin sent a message to President Woodrow Wilson. The Soviet leader asked what kind of help America could provide if Russia were to reenter the war. At the same time, Trotskii approved a landing of British troops at Murmansk to protect stored Allied supplies that had been shipped to the northern port, which had been opened in 1916 after the construction of a railroad line. There was also concern that the Germans might take over the Arctic town and convert it into a submarine base. The brief Soviet attempt at friendly relations with Russia's former allies solved Seversky's immediate problems.[4]

The young pilot had no military position, and his mother, whom his father had divorced, had little chance of surviving in a country that now was dedicated to exterminating the upper classes. Soviet Russia's pseudo lovefest with its former allies opened the door of opportunity for Seversky to travel to the United States. He sought official permission to leave Russia on the basis of his earlier appointment to the Russian naval aviation mission. In addition, he wanted to secure a newer type of prosthesis in America that would improve the function of his right leg, which was only a stump. Trotskii, soon to become commissar of war, signed the government pass for Seversky. Early in March Seversky and his mother got on a train that would take several weeks to travel 6,117 miles from the former capital to Vladivostok, on the Pacific Coast. Their rail car companions included Japanese diplomats whose embassy in Petrograd no longer existed. On the Trans-Siberian Railroad, Seversky more than once came close to being murdered by belligerent armed sailors or soldiers who entered and searched the train for anti-Bolshevik Russians. The fortunate pair survived, had a brief stay in Japan, and then boarded an oceangoing vessel for a two-week voyage that brought them to San Francisco on April 21, 1918—ten days after Safonov and his bride had flown to Finland.[5] Seversky remained in the United States, where he married, became a citizen, invented aircraft technologies, designed and manufactured military planes for the U.S. Army Air Corps, and wrote books and articles about airpower.

Unlike Seversky, Safonov and his wife, Ludmila, left Finland when the civil war ended in the summer of 1918. The key to their travels was the Treaty of Brest-Litovsk, which was signed on March 3 and approved by the Congress of Soviets a few days later. The document ended Russia's war with the Central Powers and removed the Baltic states, Finland, Poland, Ukraine, and White Russia (now Belarus) from the former empire. Except for independent Finland these territories were occupied by German forces and some Austrian troops. With his new Finnish name and official papers, Safonov and his wife managed to get across German-controlled Ukraine. He briefly joined the anti-Soviet forces in southeast Russia. The couple then went on to Persia (today's Iran) and after that to the British colony of India, where he joined the Royal Air Force. Finally, the couple moved on to China, where Safonov assisted that country in forming a naval air arm. Unfortunately, he perished in an aircraft crash; Ludmila and her two children eventually found a home in the United States.[6]

Like Igor Sikorsky, Safonov and Seversky were part of a small group of pilots who abandoned Soviet Russia early on and, sooner rather than later, ended up living outside Soviet-held territory. The majority of other pilots chose opposing sides between Reds (Communists) and Whites (anti-Communists) in the Russian Civil War. The conflict proved far more destructive to property and human life than the horrible U.S. Civil War. Some commentators argue that the internal battle began later in 1918, but other experts, such as English historian Evan Mawdsley, claim that the Russian Civil War began on November 7–8, 1917. Lenin believed that the ineffectual Provisional Government should be abolished before the Second Congress of Soviets opened so that the Bolshevik Party could play a dominant role in the formation of a new Soviet government. As it happened, Aleksandr Kerenskii, the Provisional Government's leader, inadvertently helped Lenin achieve his goal. Kerenskii took steps to weaken the party—such as closing down its official press—but he ended up activating a Bolshevik response. Under Trotskii's guidance, the insurgents had formed a Military Revolutionary Committee (MRC) that dismantled Kerenskii's initiative. The MRC then occupied the railroad station, post office, and telegraph station; liberated the Bolshevik press; and later that day took the Winter Palace headquarters of the Provisional Government and arrested most of its members. (Kerenskii had escaped from the Winter Palace earlier in an automobile provided by the U.S. Embassy.)[7]

To help engineer the MRC attack on the Winter Palace, the Bolsheviks spread the word that Kerenskii was planning to prevent the Second Congress of Soviets from meeting. When Trotskii informed the opening session of the Congress that the Provisional Government no longer existed, moderate socialists protested vehemently. As disciples of Karl Marx, they believed that the tsarist autocracy would have to travel through a fully developed capitalist-bourgeois economic stage before socialism could truly emerge. When Trotskii appeared to condemn them to the dustbin of history, the moderates withdrew from the Congress altogether. The large Socialist Revolutionary Party that represented Russian peasants then split: some joined other moderates in leaving the Congress; the extremist members remained as allies of the Bolsheviks, although they too withdrew five months later. Meanwhile, the opposition formed an all-Bolshevik cabinet, headed by Lenin, that replaced the ministers with commissars who supervised major government departments called commissariats. The Congress also approved two key measures—a Decree on Land, which abolished private ownership of land and basically endorsed ongoing peasant confiscations, and a Decree on Peace, which sought a formal end to war, and not just for Russia, but for all belligerents.[8]

There is evidence to suggest that the events of November 7–8 actually caused the Russian Civil War, and the reasons go well beyond the claim by traditional socialists that the former empire was unprepared to enter Marx's final economic stage. Just as important, many socialists—including some Bolsheviks—fully expected that the Second Congress of Soviets would be led not by one single party, but by a coalition of several socialist parties, regardless of the decisions that the Congress might reach on the issue of governance. Thus, several parties either outright opposed or at best had mixed feelings about the new Soviet government. Even the Left Socialist Revolutionaries walked out, after voting against the Brest-Litovsk Treaty at the Fourth Congress of Soviets. The anger of that party erupted when two party members murdered Wilhelm von Mirbach, the new German ambassador to Russia, and a third fired two bullets into Lenin's body. Fortunately for Lenin, his chauffeur, Stepan Gil, had enough sense to drive the wounded man to the Kremlin, where loyal physicians could be called to take care of him. Had Lenin ended up in a hospital, he might have been "treated" to a shorter life—by a Socialist Revolutionary doctor.[9]

Because Lenin could form the new government around the Bolshevik Party, he had no need for a Constituent Assembly. In the elections for the Assembly, held in mid-November 1917, the Socialist Revolutionaries held more votes and thus amassed more delegates than any other party. After the Bolsheviks forcefully closed the Assembly, the Socialist Revolutionaries took a leading role in creating the Committee of Members of the Constituent Assembly. The anti-Soviet government movement, known by its Russian acronym, Komuch, centered its activity at Samara, near the middle Volga River. Ironically, one of the cities that it occupied was Simbursk, where Lenin was born as Vladimir Ilich Ulianov. Interestingly, although the creation of a new government on November 7–8 seemed focused on Great Russian nationals, Russia actually was a polyglot of different languages and nationalities. Many of the peoples in the former Russian empire, ranging from Armenians to Ukrainians, wanted either independence as a separate country or membership as a semi-autonomous state in a federal union similar to Switzerland and the United States. Such decisions would have been left to the Constituent Assembly, but it was abolished soon after the delegates elected Socialist Revolutionary Party leader Viktor Chernov as its chairman.[10]

The Decree on Peace also angered the Allied Powers, which chose to support anti-Soviet groups. Initially, the Allies had expressed hope that the Bolsheviks could be overturned, on grounds that it might open the possibility of restoring some type of Eastern Front to force Germany to reduce its strength on the Western Front. For one reason or another, the Allied governments sent troops to the fringes of Russia. American, British, and Japanese troops entered Siberia; British and American soldiers landed in North Russia. The opening of the Black Sea after November 1918, when the Great War ended, enabled the British and French to send forces to South Russia. In 1919 the British sent increasing numbers of transport ships into the Black Sea carrying military supplies for White Armies in the southeast of European Russia. Meanwhile, the Decree on Peace served as a catalyst in marshaling the opposition of key high-ranking Russian officers who would play leadership roles in assembling military forces to fight the Soviet government. In November 1917, former Russian army chief of staff General Mikhail V. Alekseev made his way some 950 miles southeast of Petrograd to the Don Cossack capital of Novocherkassk.[11]

Alekseev would be joined by several generals who escaped from prison ten miles south of Mogilëv's Stavka headquarters. Two of them are noteworthy: General Lavr G. Kornilov, the former commander of Russian armies; and General Anton I. Denikin, former commander of Russia's Western Front. Colonel Viacheslav M. Tkachev traveled south at about the same time. Shortly after the Bolshevik Revolution, Tkachev resigned as head of Russian military aviation at Stavka. His journey took him below Novocherkassk to his homeland among Kuban Cossacks in the region around the Kuban River. There he joined a White partisan group under Colonel Kuznetsov. Later he established an airplane squadron. Eventually he flew a British D.H. 9 aircraft and became the leader of anti-Soviet aviation in South Russia, first under Denikin and then under General Petr Nikolaevich Vrangel' (often rendered incorrectly in English as Wrangel). In December 1917 the general officers began raising a military force called the Volunteer Army in Novocherkassk. As one might understand, given the circumstances, the majority of the more than four thousand men who volunteered for service had been officers in the tsarist army; they were augmented by a smaller number of noncommissioned officers.[12]

By early December 1917 the Soviet government had learned what was going on in Novocherkassk and nearby Rostov, where the Volunteer Army was being established. Lenin called on Vladimir Antonov-Ovseenko, who had led the Bolshevik conquest of the Winter Palace on November 7, to take the Red Guards—augmented by detachments of armed and loyal workers from Petrograd and Moscow—south and destroy the White movement. Together they went to Kharkov, which became the headquarters of the Soviet force. With far more men than the Volunteer Army and local Cossacks, the Soviet military took an armored train on the Voronezh–Novocherkassk Railway. The Red Guards and workers had machine guns, artillery, and even five aircraft for scouting purposes. In February, after a number of firefights, the Volunteer Army—threatened by encirclement—retreated into the Steppe region east of the Sea of Azov. Sometime later, Kornilov died when a Soviet artillery shell exploded next to his small farmhouse headquarters; a metal fragment had lodged in his head. General Denikin took command of the Volunteer Army. Through perseverance and leadership, by the fall he was able to enlarge his forces, win numerous battles against the Red Army, and gain control of the region north of the Caucasus up to the city of Stavropol'. In late 1918 Denikin was able to expand the work of the Volunteer Army, adding more men, including a large number of Cossacks. The Whites came to control the North Caucasus Region.[13]

The start of civil conflict and the uncertainty inherent in the December armistice prompted the Soviet government to worry about the possible resurrection of a military that would be loyal to the new regime. The government formed a Committee on Aviation on November 10, 1917, two days after the Bolsheviks took power. By the end of December the group had evolved into the All-Russian Board for the Administration of the Air Fleet, under the chairmanship of Konstantin V. Akashev. Lenin assigned the board the responsibility for preserving as many air units and flying schools as possible. The task was to prove difficult. Within a few weeks of the armistice, both soldiers and airmen had begun leaving their posts, and by January 1918 the Russian Army had quit fighting. As a result, the haphazard collection of willing pilots and their flyable planes had dropped to about 125 aircraft in 36 squadrons, with between 3 and 4 pilots and aircraft in each detachment. The squadrons were located primarily in or near Petrograd, Moscow, and other urban centers such as Tula, about 160 kilometers (100 miles) south of Moscow; Vitebsk, about 300 kilometers (186 miles) west of Moscow; and Saratov, about 750 kilometers (465 miles) southeast of Moscow, by the Volga River.[14]

The aircraft available to the Aviation Board in 1917—whether imported or built under license in Russia—were few in number and on the verge of being obsolescent. Existing airplanes represented British and French models from manufacturers such as Caudron, Farman, Morane-Saulnier, Nieuport, Sopwith, and SPAD. To be sure, there were a number of Russian-designed aircraft, such as the Grigorovich flying boats and Sikorsky's Il'ia Muromets planes. But it made no difference. The production of Russian-manufactured aircraft suffered from the same limitations and problems as goods made by other industries in Russia during the revolutionary period. For example, Grigorovich, the only major aircraft designer during the Great War who stayed and survived in Soviet Russia after 1917, could not resume drafting and building aircraft for another five years. Nevertheless, the Soviets operated some of his existing M-9, M-15, and M-20 flying boats on rivers after the German navy controlled the Baltic. Germans and Austrians occupied most of the northern shore of the Black Sea after the signing of the Brest-Litovsk Treaty.[15]

Soviet Russia also could not use Sevastopol' as a flight-training center; the area was overseen by a Ukrainian government led by General Pavel Skoropadsky, who collaborated with the Germans. The Aviation Board did latch on to Gatchina. Interestingly, its name changed temporarily to Trotsk in 1919 to honor War Commissar Trotskii, who took a commanding role over Red forces that prevented a White group from taking Petrograd. After Lenin's death from a fatal stroke on January 21, 1924, Trotskii lost his leadership position to Stalin (whose real name was Iosif V. Dzhugashvili). Trotskii was exiled to Kazakhstan in 1927 and then banished from the Union of Soviet Socialist Republics two years later. He was murdered in his Mexico City villa by a Soviet agent in 1940. Meanwhile, Petrograd had become Leningrad, and Trotsk had reacquired the name Gatchina. The most important of the early flight-training centers that the Aviation Board was able to revive for the Soviets were in Petrograd and Moscow. Unlike the 1910–1914 period, when Grand Duke Aleksandr favored offering pilot training exclusively to the sons of nobility, the Aviation Board avoided upper classes by encouraging workers to learn how to fly. Board members actively sought chauffeurs as flight-school candidates; they came from the lower classes, were used to working with machines, and hopefully knew something about reciprocating, fuel-powered engines.[16]

Members of the Aviation Board felt they could not trust tsarist pilots who had been born to gentry or noble families, but they often encouraged their use as instructors in flight-training programs. The pilots would be overseen by a loyal Bolshevik Party member, whose job was to make certain that the instructor did not speak against party and state. The upper-class pilot who behaved might be trusted with other duties. With justification, board members remained suspicious of pilots who were neither peasants nor workers. A case in point involved Nikolai I. Belousovich, descended from gentry, who resided near Omsk in Western Siberia. Awarded medals and the rank of staff-captain during the Great War, he remained at his post after the December armistice, and in 1918 he became one of the early tsarist pilots to serve as an instructor. His cooperation won him a slot as a pilot with the Eleventh Aviation Squadron. By summer, as the White movements expanded, Belousovich realized that he might end up in combat missions against men with whom he had served during the war, so he fled from his squadron and made his way to North Russia, where the British had troops and aircraft. On August 4, 1918, he joined the Slavic-British Aviation Corps.[17]

By contrast, Iosif S. Bashko descended from Latvian peasants in the Vitebsk Province. Because he had completed a basic program of education, he was able to attend the Vladimir Military School, from which he graduated with a commission as a second lieutenant. He then entered the Officers' Aeronautics School at Gatchina, emerging as a qualified pilot; eventually he became commander (pilot) of an EVK Il'ia Muromets reconnaissance-bomber aircraft. By 1917 Bashko had headed an EVK detachment and held the rank of lieutenant colonel. When the Germans began their unopposed offensive in February 1918, Bashko inadvertently made a good decision. He flew the four-engine plane to the city of Bobruisk, southwest of Mogilëv. Luckily for him a Polish military unit occupied the city on behalf of the Germans, and the Poles chose not to disturb him or his plane. The collaborationist Poles hoped that Germany would reunite Poland, which had been split in the eighteenth century among Austria, Germany, and Russia. In June 1918, after the Brest-Litovsk Treaty awarded Germany major parts of the former empire, including White Russia, its occupation troops approached the White Russian city of Bobruisk, forcing Bashko to seek safety by flight to Moscow at night in an Il'ia Muromets plane.[18]

During the flight two of Bashko's engines ceased working, and he was forced to crash-land his plane some 96 kilometers (60 miles) southwest of Moscow in the province of Smolensk. Local authorities thought that Bashko might be some type of spy, so they arrested him and escorted him to

Moscow. The authorities in Moscow realized that Bashko had come from a peasant background and had served as an EVK pilot and detachment leader. Moreover, since Bashko was a Latvian who supported the Soviet government and was willing to fly for it, the locals abandoned all thoughts of treating him like a spy or criminal. Before long he became the commander of the Red Squadron—the name given to the remaining fragment of the former EVK. However, he still had to confront the same shortcomings and limitations faced by all other Soviet pilots and flight commanders. There was a severe shortage of mechanics—especially engine mechanics. Skilled workers, adequate machinery and tools, and aircraft spare parts all were virtually unavailable. Lubricants and gasoline basically did not exist. At first the Caspian oil town of Baku was in Soviet hands, but all twenty-six Bolshevik leaders in Baku were shot by anti-Soviet Russians. As a result, Soviet aircraft had to be powered by ersatz fuel such as alcohol, which burned badly, reduced the performance of the aircraft, caused engine failure, and produced fumes during flight that sickened pilots or gave them headaches.[19]

During 1918 Soviet authorities at least partially addressed the serious problems that their pilots were experiencing in flying, maintaining, and augmenting aircraft. In June aircraft factories were nationalized in an effort to restore some manufacturing of airplane parts and, eventually, of airplanes. As that took place, the Central Aviation Park in Moscow became the main repair station for army aircraft. With civil war dragging on, additional aviation repair parks were established in cities such as Nizhnii Novgorod, Petrograd, Smolensk, Tver', and Iaroslavl'. Over time every supply train behind each Soviet army contained cars that served as aircraft repair shops. In May 1918 the Aviation Board was replaced by the Worker and Peasant Red Military Air Fleet, which was linked to the Worker and Peasant Red Army that had been founded as a volunteer unit in January 1918 and then expanded by conscription in May. Near year's end Lenin encouraged Nikolai E. Zhukovskii to establish the Central Aero-Hydrodynamic Institute, known best by its Russian acronym, TsAGI. In TsAGI's efforts to research and develop better aircraft, the institute hosted a cadre of younger but promising experts in aviation design, including Aleksandr A. Arkhangelskii, Konstantin A. Kalinin, Aleksandr A. Mikulin, and Andrei N. Tupolev, and senior designer Dimitrii Grigorovich.[20]

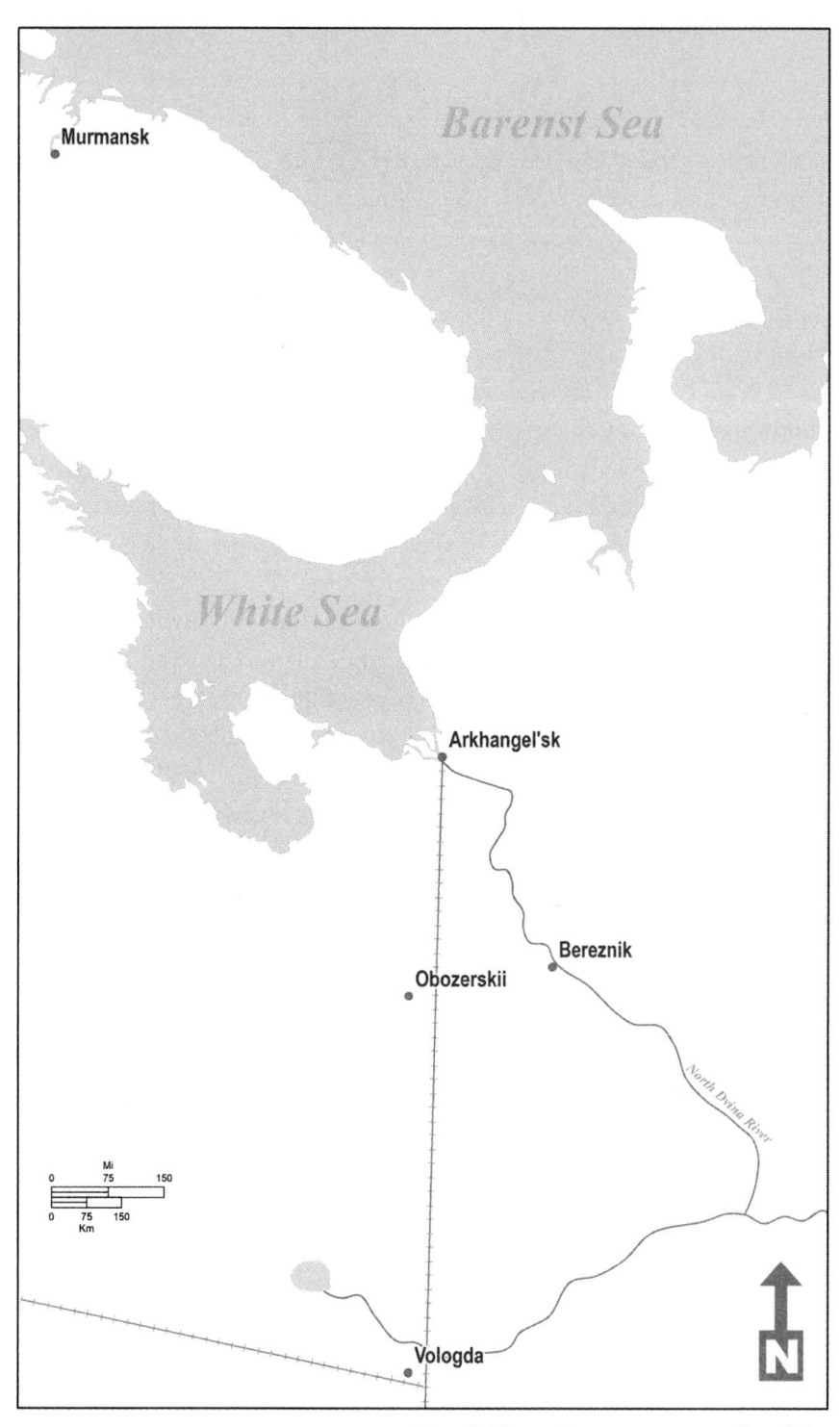

Map 5. North Russia during the Civil War

Map 6. From Omsk, Siberia, to Ufa during the Civil War

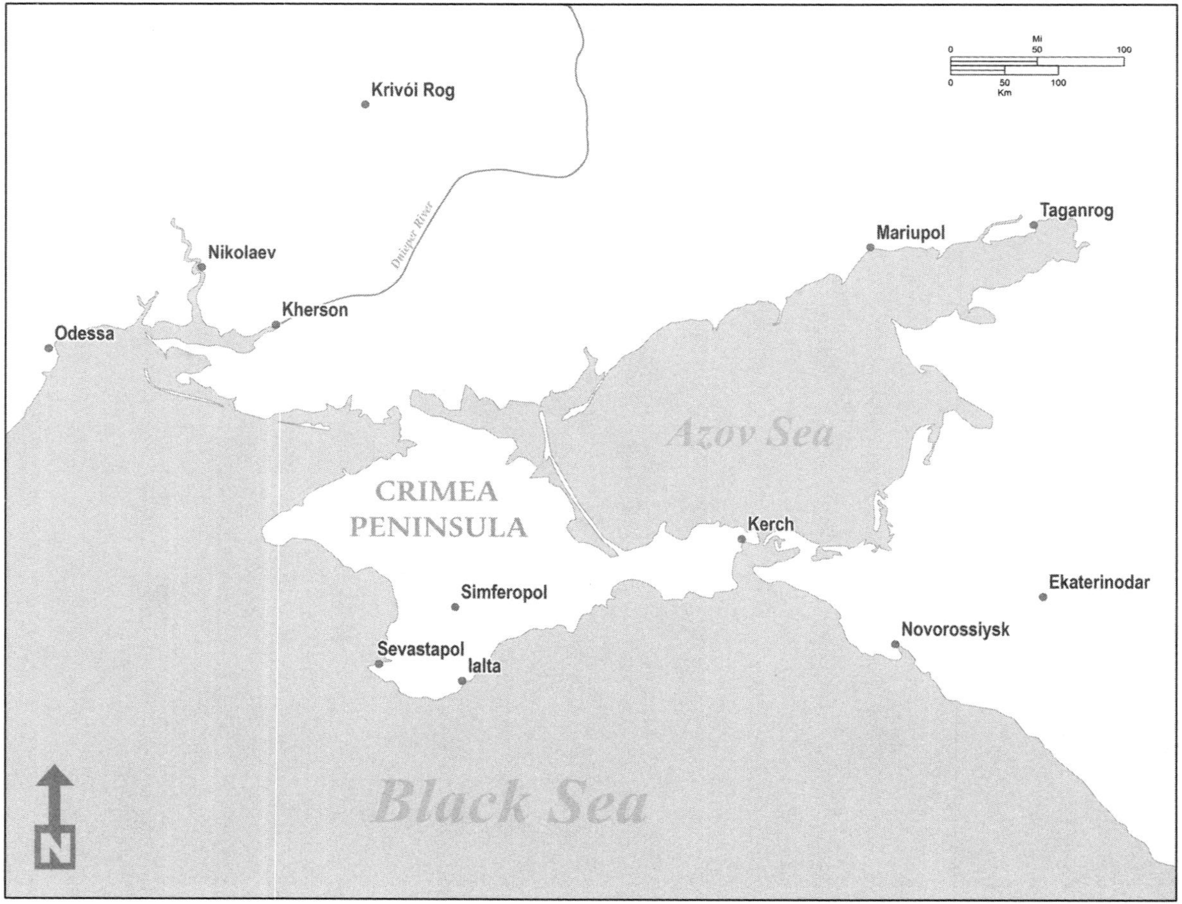

MAP 7. Crimea during the Civil War

As the Soviet government began to address the obvious deficiencies in military aviation and at the same time fought the White Volunteer Army in Southeast Russia, it confronted another equally troublesome problem—the Ukraine. The Ukraine was not a White area that wanted to contest and compete with the Bolsheviks for control over Russia. Rather, the jurisdiction's thirty-two million persons, who spoke an eastern Slavic dialect different from the Russian language, wanted separation. What made the situation especially interesting was that in urban areas such as Kiev, a substantial minority of Russians often constituted a majority of the population. And Kiev, of course, was Russia's first capital city. Ukrainian peasants dominated rural sections, which helps to explain why Socialist Revolutionaries played such a prominent role in the Rada, the Ukrainian parliament, which declared the Ukraine an independent country in January 1918. History and the northern need for Ukrainian grain convinced the Bolsheviks that Russia would not be whole without the Ukraine. The result was that Red Guards and the First Revolutionary Army moved into the rebellious territory by armored trains.[21]

The train carrying the First Revolutionary Army rode into southern Ukraine under the leadership of Aleksandr V. Polupanov. The force would have the assistance of a remnant of the Sixth Corps Aviation Squadron under the command of its former captain, Lev K. Grinshteii, a graduate of the Sevastopol' Aviation School. Grinshteii was born to a humble family and raised in the Ukrainian city of Vinnitsa. In February Grinshteii's squadron did some scouting for the First Revolutionary Army in its successful battle against Rada troops near the town of Zhmerinka, about 50 kilometers (31 miles) south of Vinnitsa. Victory there prompted Mikhail A. Murav'ev, who commanded the Red Guards that took and occupied Kiev, to appoint Grinshteii as his commissar of aviation operations. Rada ministers, however, quickly undermined the Soviet victory. The ministers fled from Kiev to the west and negotiated a separate peace with the Central Powers at Brest-Litovsk. Three weeks after the Red Guards conquered Kiev, German troops drove the Bolsheviks out of the city and, aided by Austrian soldiers, soon seized all of the Ukraine. Not long after that, the puppet (and German-friendly) regime of General Skoropadsky replaced the Rada.[22]

Thanks to the success of Rada ministers in hammering out a peace agreement with Central Powers, the occupation of the Ukraine began before the Germans approved it on March 22, 1918. The Brest-Litovsk Treaty, ratified six days earlier by the special Fourth Congress of Soviets, made Soviet Russia's loss of the Ukraine official. At the end of the same month, Czech-Slovak leader Tomáš G. Masaryk secured an agreement with Lenin's Soviet government that permitted Czech and Slovak soldiers who had primarily been prisoners of war (POWs) to leave Soviet Russia at Vladivostok via the Trans-Siberian Railroad. Masaryk and important colleagues such as Eduard Beneš wanted the soldiers in France and on the Western Front, in hopes that an Allied victory—with the aid of Czech and Slovak participation—most likely would lead to future Allied support for the creation of a new country, Czechoslovakia, drawn from significant portions of a defeated Austro-Hungarian Empire. Indeed, hatred for Austria already had inspired some Czechs and Slovaks, who had settled years earlier in Russia, to establish a Czech Legion of soldiers, which fought alongside Russian troops against Austria from 1914 to 1916.[23]

In 1917 the Provisional Government opened the POW camps to Czech recruiting officers. By the time of the Bolshevik Revolution the Czech Legion had grown from a brigade of a thousand men to an army corps of more than 35,000 soldiers. The Germans and Austrians were extremely upset that the Soviets had decided to send the Czech Corps east to the Pacific Ocean, where Allied ships could greet the soldiers and transport them to France for duty on the Western Front. On May 25, as various trains carried elements of the Czech Corps across Siberia, War Minister Trotskii sent a telegram to urban centers along the Trans-Siberian Railway ordering local Soviets to disarm or shoot Czech soldiers. Discovery of this telegram made Czech soldiers feel betrayed by the Soviet government. The Czechs refused the order and instead fought local Reds, seized towns and cities, and took over the Trans-Siberian Railway. It was an extraordinary moment in Russian history. The process revealed the Soviets' weakness in Siberia, elevated the Whites to power in the region, and suggested that armed and organized groups could defeat the Reds.[24]

The Czech Corps' dominance over the Soviets in Siberia served as a catalyst for the emergence of a White movement in that area and the encouragement of similar groups in Southeast Russia. Challenges from these two large areas triggered a major response that severely reduced Moscow's presence in North Russia. Astonishingly, the Soviet government invited Great Britain to provide security around Murmansk. Meanwhile, following the March transfer of the Soviet capital from Petrograd to Moscow, British and Allied embassies—including that of the United States—moved to Vologda. About 500 kilometers (310 miles) east-southeast of Petrograd, Vologda sat at the junction of rail lines running north to Arkhangel'sk and east to Vladivostok. Embassies were located there purposely to give embassy personnel two options for escape if they needed to flee Soviet Russia. In contrast to the embassies, the substantial British, French, and Italian military missions moved to Moscow with the Soviet government. Even after the Brest-Litovsk Treaty had been approved, Allied military officers hoped that the dreadful terms of that document might persuade Soviet Russia to resume war with Germany.[25]

The Allied alternative to a Soviet change of heart about fighting Germans involved supporting both the Czech Corps and White groups, which they thought together might form a force large enough to challenge the Germans—at least in their occupation of Russian territory—and prompt the German Army to respond with soldiers taken from the Western Front. In June 1918, as part of this process, the British military mission in Moscow warmly welcomed several Russian aviators, who were led by Russia's renowned ace fighter pilot, Aleksandr A. Kozakov. Among others, the men included Sergei K. Modrakh, Aleksandr N. Sveshnikov, and Sergei K. Shebalin. All of them had held the rank of staff-captain or higher in the pre-Soviet air force, earned numerous medals for bravery in combat missions, belonged to the Russian Orthodox Church, came from wealthy or noble families, and were persona non grata in Soviet Russia. The British military recruited the group to go to Murmansk in North Russia and join British security officers and troops. Originally, the British thought that the Russian pilots would be part of a larger force of American, British, some French, and White soldiers that would move south and penetrate the interior of Soviet Russia as they merged with other White and Czech Corps troops.[26]

It never happened, for a spate of reasons. The French had no more than a token force in North Russia, and on July 17, 1918, President Wilson sent a qualifying memorandum to the Allied governments in which he made clear that even though the 339th Infantry Regiment would be under British control, its 4,500 American soldiers were assigned only to help protect military stores. Wilson said he would withdraw the regiment if the Allies intended to use the troops for some purpose other than the defense of supplies. Also, White forces initially were small and poorly trained. Finally, North Russians and Siberian forces were simply too far apart ever to merge. While these flawed Allied plans simmered, the Russian pilots in Murmansk enlisted as lieutenants in the Royal Air Force and joined the Slavic-British Aviation Corps. On August 1, 1918, a flotilla of British ships brought soldiers and pilots to Arkhangel'sk. Two weeks later the First Slavic-British Aviation Squadron was formed. The unit initially operated below Arkhangel'sk at Obozerskii, but on September 17 it set up a more permanent base and aerodrome near Bereznik.[27]

Bereznik was located a little more than halfway between Arkhangel'sk and the Red Army, which was about 130 kilometers (81 miles) south of the port. This squadron and others that came later generally operated some Nieuports that had been shipped to Murmansk but not sent to Russia's interior. Most squadrons flew British aircraft such as the R.E. 8 and D.H. 9. The Reconnaissance Experimental #8 came from the Royal Aircraft Factory, which was run by the British government. A two-place biplane, it provided the observer with access to a mounted Lewis machine gun to protect the plane from enemy aircraft. Obviously intended for scouting duties, the R.E. 8 could also carry a bomb load of more than 300 pounds. Geoffrey de Havilland designed the D.H. 9, which was built by the Aircraft Company, Ltd., commonly referred to as Airco. Manufacturer George Thomas established the firm in 1911. An advanced version of the famous D.H. 4, the D.H. 9 proved to be a very successful and long-serving military aircraft. Depending on whether it was equipped with a Fiat A-12 or a 430-hp Napier Lion engine, the plane could fly somewhere between 118 mph and 144 mph. The aircraft was protected by one or two machine guns on a Scarff Ring, manned by the observer, along with a forward-firing Vickers machine gun operated by the pilot. It also could carry a practical bomb load of 350 pounds.[28]

By late June 1918 the Soviet government realized that British protection of Murmansk had evolved into a form of intervention that favored anti-Soviet Russians. At first, however, the Worker and Peasant Red Army and its air fleet failed to send troops and airplanes to challenge the British and later the American military because of the unexpected revival of the White movement, thanks to the Czech Corps. At the beginning of August, when Soviets learned that British ships with Allied troops and planes had landed in Arkhangel'sk, Lenin took action. The Red Sixth Army was assigned to North Russia, and Lenin then telephoned the leaders of the Red Air Force, Eduard M. Skliansk and Aleksei V. Sergeev. They disclosed that there were two aircraft squadrons near Vologda, but said they would have to be moved farther north to be able to conduct missions. That meant locating, preparing, and supplying an aerodrome south of Arkhangel'sk.[29]

In one sense, the number of Soviet ground and air forces could only create mischief rather than defeat and evict British and American troops from North Russia. During the Great War, a typical Russian army might have 2 corps with at least 2 divisions in each corps. By contrast, the Red Sixth Army only had 9,000 men initially and eventually just 18,000 soldiers. It would be generous to suggest that the army was composed of 2 full divisions. It only had 34 pieces of artillery, and the guns and troops had to be spread over a vast amount of territory. The 2 aircraft squadrons had enough planes for 1 squadron with 1 or 2 backup planes. As for Soviet pilots, the one who received the most attention was Sergei F. Smirnov. He was noted not so much for his piloting skills as for the fact that he cherished communism. He and his comrades flew Caudron G. 4 biplanes and occasionally a Russian-built Nieuport. The air unit also had a balloon. Gaston Caudron designed the twin-engine G. 4 and, with his brother René, founded the Caudron Company that built the plane in France. Powered by Le Rhône motors, the G. 4 had a speed of 120 kmh (77 mph). Even with a machine gun, the pilot and observer dared not engage the British D.H. 9.[30]

Although the G. 4 could carry up to 113 kilograms (250 pounds) of bombs, Red Army pilots had artillery shells, not bombs, which they dropped with little success at hitting a target. Aside from scouting, many of their missions involved distributing propaganda sheets to villages that were under White-British control. The hope was to garner support for the Soviet government and opposition to Nikolai V. Chaikovskii, who headed the Supreme Administration of the North. The effort suffered interference from the British, and struggles among Whites between those who favored a conservative administration and those preferring a socialist one. During such squabbles the First Slavic-British Aviation Squadron played a key role in a victory over a portion of the Red Sixth Army. In October in the region of Sel'tsa along the railway line between Vologda and Arkhangel'sk, reconnaissance flights discovered a weak sector that enabled the British to surround their opponents. As winter approached, both sides virtually ceased their flights because of the frigid cold and abundant snow. Occasionally, in perfect weather, a scouting mission would be conducted. For example, on December 24, 1918, Lieutenant Sveshnikov completed a reconnaissance mission designed to scout Red Army winter positions. Eight *versts* (five miles) from the aerodrome at Bereznik, the engine of his Nieuport caught fire, and he had to descend into the woods. He actually survived the crash landing, but because of his injuries he could not extract himself from the aircraft wreckage. As a result, he froze to death.[31]

CHAPTER 9

AVIATION AND THE CIVIL WAR

★ ★ ★ ★ ★

A MONTH LATER, IN JANUARY 1919, ace pilot Aleksandr Kozakov almost joined Aleksandr Sveshnikov in death. Kozakov flew a combination scouting and bombing mission over the winter position of the Sixth Red Army. Apparently he dropped his altitude to improve his chances of hitting the target that he had selected. He was flying low enough that small-arms fire hit his plane. One of the bullets struck his chest at an acute angle, passed through a corner of his lung, and escaped out his shoulder. Although he was bleeding and in pain, he nevertheless made it safely back to the Bereznik Aerodrome and the care of qualified medical personnel. By the time he fully recovered in March his wound and combat service had earned him the British Distinguished Flying Cross. When Kozakov returned to duty he tried unsuccessfully to train a couple of White Russians as pilots. Despite the setback there were enough veteran Russian pilots to create a second Slavic-British Aviation Squadron, headed by Nikolai I. Belousovich. Along with three other squadrons manned only by English pilots, North Russian aviators clearly enjoyed air superiority over the two Soviet squadrons.[1]

The Soviet squadrons rarely had more than ten or eleven aircraft. The British imported additional planes, including D.H. 9s and Sopwith Snipes, forcing Soviet opponents to face up to nine times as many aircraft as they had in their own units. The Snipe, which went into production in England in 1918, was built around a 230-hp Bentley Rotary (B.R. 2) engine, and was the ultimate development of the small rotary-powered aircraft produced near the end of the Great War. The plane's 121-mph speed made it the best fighter-scout plane of the war. It carried four light bombs and on the fuselage in front of the pilot there were two Vickers machine guns that fired synchronically through the propeller blades. The performance of the aircraft that Soviet squadrons used in North Russia never matched the D.H. 9 and Sopwith Snipe. On the other hand, the Soviet government nationalized the Dukh factory in Moscow in the spring; the plant gathered workers and materials in the second half of 1918, and it began building airplanes before year's end.[2]

The revived and Soviet-controlled Dukh factory benefited from the former Lebedev Aeronautics Company. In 1917 Dukh acquired a license from England to build the Sopwith 1 ½ Strutter. Designed by the Fairey Aircraft Company, the aircraft prototype went through a series of successful tests in December 1915 and went into production, early in 1916. The Strutter was the first British airplane to be equipped with an interrupter gear that enabled a fixed machine gun to fire through the rotating propeller. The outer struts, supporting the top wing with the center section of the fuselage, were half the size of the outer interplane struts, giving it the moniker "1 ½ Strutter." The biplane's nickname became its normal designation. The truly impressive feature of the aircraft was its built-in triple functions: it could be a two-place scout, with a machine gun mounted on a Scarff Ring for the observer; a single-seat fighter; or a single-seat attack plane, carrying 224 pounds of bombs. With a 130-hp engine, the Sopwith 1 ½ Strutter had a maximum speed of 106 mph.[3]

Early in 1916, the aircraft represented front-line technology. By year's end, however, German planes such as the Albatros D. 1 and D. 2 had made the Sopwith 1 ½ Strutter obsolete. Accordingly, the British turned the aircraft over to their navy and gave the Russians a license to build it. In the second half of 1918, some equipment and templates and a few workers from the former Lebedev plant moved from Petrograd to Moscow, and to the renamed Dukh facility, now called GAZ-1, for *Gosudarstvennyi Aviatsionnyi*

Zavod-1 (State Aviation Factory No. 1). It was a struggle to get supplies and materials needed to build an airplane. There were few qualified aviation workers. Some had become mechanics, voluntary or conscripted, in the Soviet military; others had died from health complications caused by the widespread famine in urban centers. Nevertheless, between 1918 and 1923 GAZ-1 managed to put together more than a hundred Sopwith 1 ½ Strutters. The plane became one of the standard aircraft for the Soviet military during this early period of post-revolutionary Russia.[4]

Meanwhile, by early 1919 the Worker-Peasant Red Military Air Force (which this book will call the Red Air Force) had about 350 aircraft. The planes were old, used, and obsolete, and involved several different aircraft models, which made repairs extremely difficult. Many of the planes could only fly one or two missions with ersatz fuel before requiring serious attention from a knowledgeable mechanic, who in turn might not have the needed replacement parts. While the Red Air Force used rolling train cars as workshops, with mechanics and repair materials, there were only a few railroad lines away from the central regions. Moreover, the Whites could interrupt train travel by damaging or removing rails. Still, except in North Russia, the Reds always had more aircraft than the Whites—on the Eastern, Western, and Southern Fronts. One key to the Soviet victory in the civil war was that the Reds always held the Russian heartland, which comprised urban centers, industries, and more railroad lines than other sections of the country. True, food was scarce and some raw materials were difficult to secure, but the Reds had an abundance of population, hardware, and troops, which the Whites lacked, since they were primarily deployed around the periphery of the heartland.[5]

Because the Whites approached the heartland in smaller units, and from separate peripheral areas, in 1919 they were vulnerable to easy defeat by the much larger Red Army and Air Force. Moreover, the Whites failed to conduct a well-coordinated, carefully timed offensive against the Moscow-based Soviet government. Technically, the various White forces did have an overall leader, but distance and communication problems among anti-Soviet groups made any type of collaboration extremely difficult. To top that off, the head of all the White ground forces turned out to be a navy admiral, Aleksandr V. Kolchak. In late summer 1918 Kolchak ended up in

Japan, where he had several talks with General Sir Alfred W. F. Knox. The British officer had served his country as military attaché in Russia during the Great War. In October, when he took a special train from Vladivostok to Omsk, Siberia, his conversation companion—in a discussion of the future of the anti-Soviet movement—was Admiral Kolchak.[6]

On November 4, 1918, at the urging of General Knox, the ruling directorate of the White government in Omsk ended up with Kolchak as the new minister of war. Two weeks later, a coup against the directorate, supported by a contingent of British soldiers equipped with machine guns, helped install Kolchak as the dictator of the Omsk government. Directorate members were humanely allowed to flee the area as Kolchak simply took over the existing system of ministries and personnel of the government. Led by the British, the Allied powers—with the exception of the United States—recognized Kolchak as the official head of the White movement. He adopted several titles, including Supreme Ruler (*verkhovnyi pravitel'*). General Knox commanded the British Military Mission in Siberia. Along with England, the United States sent some military hardware and supplies to Kolchak via Vladivostok, but the Wilson administration, leery about the lack of any real coordination among the Whites, chose not to recognize Kolchak as the top commander over all Soviet opponents; the endorsements from widely separated White groups acknowledging Kolchak's authority were more honorary than realistic.[7]

Even before the Red Army and Red Air Force challenged Kolchak's forces along the Eastern Front in the warmer months of 1919, the Allies in North Russia confronted serious challenges. Morale among American, British, and French troops plunged dangerously. The Great War had ended on November 11, 1918, and many of the Allied military men could not understand why they were fighting the Russians, who were a former ally. Near the end of January 1919 an American unit suffered heavy casualties, which led the Red Army to occupy the town of Shenkursk, about 50 miles south of the Slavic-British Aviation Squadron aerodrome at Bereznik. For the British, it meant that there would be no chance of linking up with White troops from Siberia. It would be impossible even to supply the Kolchak forces from North Russia. Worse yet, Allied soldiers were the first to recognize that most Russians in the north were sympathetic to the Bolsheviks (renamed the Communists) and their Soviet government. General Evgenii K. Miller, commander in chief of White forces, had to face the issue. His 16,000 unenthusiastic soldiers could best be described as mutinous. By April a battalion and an entire regiment had abandoned Miller for the Red Army.[8]

On top of the morale and mutiny problems of Allied and White soldiers, the Red Army took steps that accelerated the Allied withdrawal. In March 1919 Soviet military forces began to close the land connection between Murmansk and Arkhangel'sk. The cities were 350 miles apart by air—500 miles by ship. Unlike Murmansk, a port that was located on the open Barents Sea, Arkhangel'sk bordered the White Sea, which was only an inlet of the larger body and was mostly surrounded by land. Several freshwater rivers in that area flow into the sea and reduce the saline levels. In the north climate, the White Sea freezes over from November to May. To avoid entrapment, the allied soldiers had to be removed on vessels during the summer. Even so, Kozakov and his Slavic-British Aviation Squadron continued to fly reconnaissance and bombing missions all year. In May, for instance, his squadron dropped bombs and swept the aerodrome field of the Red Air Force with machine-gun fire; official Soviet Russian military records describe the aerodrome's destruction (*razrushenie*). In July, as Allied forces prepared to leave Arkhangel'sk, the British invited Kozakov to move to England, but he chose to stay in Russia. Reputedly despondent, Kozakov later took off in a Sopwith Snipe. Barely off the ground, the expert pilot abruptly flew straight up in the air. Naturally, the Sopwith Snipe stalled and fell like a rock in a crash that killed Kozakov.[9]

On August 1, three days after Kozakov's death, there was a large funeral procession, led by two Orthodox priests and followed by dozens of pilots, soldiers, and citizens of Bereznik. In his casket, Kozakov was laid to rest in a grave near the aerodrome. Allied personnel were transported out of Arkhangel'sk before the end of the month, but several White Russian pilots did not withdraw with the British. Like Nikolai I. Belousovich and Sergei K. Modrakh, they managed to make their way southeast to the White forces under Admiral Kolchak. Belousovich ended up as a flight instructor in the Military Aviation School. Modrakh was named chief of field administration for aviation in Kolchak's army.

After Kolchak's defeat, the two pilots went different ways. Modrakh moved west after Kolchak's demise and joined the White forces under General Petr N. Vrangel'.[10] Belousovich fled Russia by crossing into China, eventually taking a ship across the Pacific Ocean, and landing on the West Coast of the United States. In America, the former Russian staff-captain and British lieutenant worked as a taxi driver. He died in San Francisco on March 17, 1956.

Months before the North Russian White pilots found their way to Omsk, the Kolchak regime had found a new source for securing aircraft—the new country of Czechoslovakia, which came into being during the fall and winter of 1918–1919. The event prompted the Czech Legion to abandon the Trans-Siberian Railroad in exchange for Allied ships in Vladivostok, which would carry Czech and Slovak soldiers to Europe, where they could travel to their new homeland. As a result, the Kolchak military inherited the few airplanes that had belonged to the Czech Legion. The aircraft included a very small number of heavily used Russian biplanes built by the Anatra Aircraft Company before the 1917 Revolution. In addition, the United States supplied the Legion with some 20 D.H. 4s that had been produced under license in America. (As indicated before, the British D.H. 4 was the model for the modified D.H. 9.) In May 1919 the Allied Powers (with the exception of the United States) decided to provide the Kolchak regime with a substantial amount of war materiel. Considering the distances, those military supplies would not arrive in the region until summer, and then would have to be shipped thousands of miles more to the fighting front—on unreliable trains that no longer were protected by the Czech Legion.[11]

Earlier, the Czech Legion inadvertently provided another benefit to Kolchak that strengthened and enlarged his military forces. The original Directorate in Omsk that led the Provisional All-Russian Government had been protected by the Czech Legion. However, its five-member Directorate included two representatives of the Socialist Revolutionary Party, Nikolai D. Avksentiev and Vladimir M. Zenzinov—a composition that made former tsarist officers, many of whom belonged to the nobility, reluctant to join the Directorate's armed forces. When the departing Czech Legion removed its support and the Directorate was replaced by the military-conservative

Kolchak dictatorship, it attracted more officers to the armed forces. As occurred in White groups during 1918, many officers were willing to ignore their former military ranks, serving in essence as privates or sergeants. Indeed, most former Directorate officers who actually functioned as White officers performed duties that were well below their tsarist ranks. The influx made an impact. By late spring 1919 Kolchak's army had nearly doubled, to 125,000 men.[12]

While the White Army expanded, however, its aviation component did not. Only 3 or 4 squadrons flew with the White armed forces, and they were assigned to provide intelligence on the disposition of the Second, Third, Fourth, and Fifth Red armies, spread out over a front of 700 miles. A couple of additional aircraft may have arrived in Vladivostok in the summer, but the Whites' key offensive took place in the late winter of 1918 and the early spring of 1919. Kolchak and the British Military Mission talked about establishing a series of aerodromes across the 3,000 miles from Vladivostok to Omsk as a quicker and safer way of transporting aircraft to the White forces, but those discussions never resulted in any actions. As mentioned above, few if any aircraft arrived as part of the 97,000 tons of military supplies that the British sent to Siberia. The Omsk regime had to rely on the slow and irregular trains to transport the Allied war materiel.

Unfortunately for the Whites, the Trans-Siberian Railroad also became the victim of the independent ataman of Siberian Cossacks, Grigorii M. Semenov, who occasionally appropriated goods from railroad traffic that had been intended for Kolchak. The good news for the Kolchak military was the taking of Perm—on what the Soviets called the Eastern Front—on Christmas Day of 1918. The city, which served as the Russian center of mining in the Ural Mountains, also was the home of metallurgical plants, specifically the Motovilikha Artillery Works. Among other goods, Perm supplied Kolchak forces with 43,000 tons of coal, 350,000 tons of manufactured metals, 250 machine guns, 10,000 artillery shells, and 10 million rifle cartridges.[13]

These military goods helped the White Western Army, under General Mikhail V. Khanzhin, to begin a major offensive in the center of the front on March 4, 1919. The attack forces followed the east–west railway from Cheliabinsk in the Ural Mountains to the Volga River. Covered with snow, the terrain enabled the army to move rapidly on horse-drawn sledges. The troops captured the major city of Ufa and, after covering 250 miles in less than eight weeks, reached the town of Chistopol' on the Kama River, which fed into the Volga. Khanzhin's right flank was protected by General Rudolf Gaida and the Siberian Army, which had moved west from Perm to Viatka, a less-spectacular distance of about 90 miles. Lenin was shocked and very concerned by the White victories, which he realized were threatening the center of Soviet power. War Commissar Trotskii responded during the White offensive by adding the Turkestan Army and the First Red Army to the array of troops that already were available to the military commander of the Soviet Eastern Front, Mikhail V. Frunze.[14] (Frunze held no military rank—typical of the early Soviet period, where authorities eschewed ranks as too Western.)

The additional forces enabled the Red Army to begin a counteroffensive against the Whites. Initially, the Soviets had labeled the military action the Ufimsk Operation because it was designed to push Kolchak's soldiers back to the city of Ufa. The Red Air Force mustered about 100 aircraft and 88 pilots to assist the ground troops. Four to 6 pilots flew in each of the 17 squadrons. The field administrator for aviation on the Soviet Eastern Front was Viacheslav S. Rutkovskii. Like most of the pilots whom he commanded, Rutkovskii was not a member of the nobility. After finishing the Mstislav City School in the Mogilëv Province, he joined the Russian Army as a private and gained officer status by attending the Vilens Military School. Subsequently he completed the Theoretical Aviation Course at the Saint Petersburg Polytechnic Institute and graduated from Gatchina's Officers' Aeronautics School as a pilot in 1913. Heavily decorated for his aviation combat missions and administrative duties during the Great War, by 1917 he had earned the rank of lieutenant colonel and commanded the Tenth Aviation Division at a time when theoretically the Provisional Government ruled Russia.[15] Rutkovskii accepted the results of the Bolshevik Revolution and loyally supported the new Soviet government. He became an aviation member of an early Red Army unit in the Tula Province south of Moscow, and on July 23, 1918, he joined the staff of War Commissar Trotskii.

On the Eastern Front in 1919, Leonid A. Kul'tin was Rutkovskii's assistant field administrator. Son of a machinist, he completed a "real"

(*real'nyi*) school in 1909 before attending the Vladimir Military School. As a lieutenant, he later transferred to aviation and graduated from the flight-training program at the Sevastopol' Aviation School—just at the time that conflict began between the Allies and the Central Powers. Like Rutkovskii, Kul'tin performed well during the Great War both as a combat pilot and aviation commander. By 1917 he had earned the rank of captain. He headed the aviation defense of Petrograd after the Bolshevik Revolution and later established aircraft communication flights between Petrograd and the new capital city of Moscow. Early in the Russian Civil War of 1918 he served as assistant field administrator for aviation in the southeast; the following year he moved to the same position on the Eastern Front.[16]

Although the Red Air Force on the Eastern Front had a few of the recently built Sopwith 1 ½ Strutters, it had to perform a variety of aviation tasks flying mainly old and heavily used aircraft. It would be likely that on any given day no more than forty or so planes could be made ready to take off for a mission. Reconnaissance continued to be the most active and important function in support of ground soldiers when, on April 28, 1919, the Red Army initiated a counteroffensive against White forces. Fortunately, the surprising—and relatively rapid—triumph of the operation included the capture of many formerly White-owned aircraft that were absorbed into the Red Air Force. Because the Red Army offensive succeeded, the front became fluid and it required several scouting missions each day to upgrade the information on White movements. Once the Whites had retreated, the Red Air Force could divert more aircraft with machine guns and light bombs to attack columns of troops and destroy horse-drawn supply wagons. Each time the Whites set up a defensive line to interrupt the Red Army advance, aircraft scouts helped Red Army artillery batteries improve their fire.[17]

By early June, the Red Army and Red Air Force approached the Belaia River, 20 kilometers (about 12.5 miles) north of Ufa. The artillery shells and machine-gun fire forced Kolchak's soldiers to move south into the city. Over the night of June 7–8, the Red Army's Twenty-Fifth Rifle Division, under the command of Vasilii I. Chapaev, crossed the river and marched toward Ufa. During the day, the Red Air Force flew multiple missions over city streets, inflicting injury or death on White troops from bombs or machine guns. The Whites began fleeing the city in panic, throwing down their rifles and anything they were carrying. By June 9 Chapaev's division controlled Ufa. The division belonged to the Red Fifth Army, commanded by Mikhail N. Tukhachevskii. The attack led Frunze's Eastern Army Group in breaching and crossing the Ural Mountains and into the plains of Siberia. A month after its victory at Ufa, the same army occupied the key center of Zlatoust, which opened the way to Cheliabinsk and Omsk. At this point the Red Air force had conducted eight hundred sorties and dropped close to ten thousand pounds of bombs. The most valuable aircraft asset, however, continued to be the scouting intelligence that was gathered on the location and size of White military groups.[18]

The White forces lost an amazingly high number of men, either from injuries, death, or capture, or from soldiers fleeing to join the Red Army. By June the combat strength of Kolchak's troops had fallen to fifteen thousand men. But the Red Army also faced problems, which began soon after the conquest of Ufa. Disagreements broke out in high places within the Soviet government over the next step that the military should take. Lenin wanted to eliminate Kolchak, the assumed leader of the entire White movement, along with his government and armed forces. By contrast, Trotskii hoped to hold off an attack against Kolchak until the spring of 1920; the war commissar expected the bulk of the Eastern Front armies to move south against General Anton I. Denikin. Trotskii, who came close to resigning over the issue, was absolutely correct in his strategic thinking; by August 1919 Denikin's Armed Forces of South Russia had succeeded in moving north toward the Soviet heartland. As the Southern Front took more men and war materiel, supplies and troop replacements on the Eastern Front dwindled. Moreover, military units on the Eastern Front—including the entire Third Red Army—had to stay in place to occupy and control territory that had been won from the Whites.[19]

The active remnants of the Red Army Eastern Front pushed on. In the fall of 1919 it conquered Cheliabinsk and Omsk. While White government ministers and personnel escaped by train to the east, their leader had not accompanied them. Kolchak's tardy departure led to his eventual capture by the Communist-dominated Military-Revolutionary Committee of Irkutsk. He was interrogated for several days in January and executed early in February 1920. Lenin did not actually want Kolchak killed, because he feared it would strengthen the White resistance, so he tried to keep the news that Kolchak had been shot a secret. The elimination of the Whites' supreme leader prompted Lenin to order War Commissar Trotskii to withdraw as many troops and trains as possible from Siberia and send them to the west. As a result, Soviet power in Siberia temporarily ended at Lake Baikal some 1,500 miles from Vladivostok. Lenin had two reasons for requiring troops. First, it had become clear that the new country of Poland had rejected the Curzon Line as its eastern boundary with Soviet Russia. Accepted by the Allies on December 8, 1919, the ethnicity- and language-based border had been suggested by British foreign secretary George N. Curzon. The Poles, though, had ambitions to regain the centuries-old scope of the long-deceased Kingdom of Poland, which included virtually all of White Russia and about half of the Ukraine.[20]

Second, Lenin was really concerned that Kolchak's demise would revive General Denikin's Armed Forces of South Russia. The Soviet leader had good reason to want to get rid of Denikin and his troops. On July 3, 1919, the White general had issued Order No. 08878, described as his "Moscow Directive," which required commanders of the Armed Forces of South Russia, composed of 200,000 men, to march north along rail lines leading to Moscow, which Denikin expected his troops to occupy. Kolchak's troops had already endured a series of defeats that exposed the complete lack of coordination among separate White groups. The good news, however, was the decision by the British war cabinet on November 4, 1918, to provide Denikin with substantial amounts of military arms and equipment. On October 30, 1918, as the end of the Great War approached, Turkey had signed an armistice of surrender that opened the Black Sea, which served as a direct link to Denikin. Beginning in March 1919, the first of 45 British transport ships arrived. From those vessels Denikin's armies received 198,000 rifles, 6,200 machine guns, 500,000,000 rounds of small-arms ammunition; 1,121 artillery pieces, with 1,900,000 shells; 460,000 greatcoats; 645,000 pairs of boots; and 168 aircraft.[21]

Although the British had supplied 168 aircraft, the actual number of planes in the inventory of the Armed Forces of South Russia is open to debate; it varied with time and use. The Whites had added a number of

German planes to their stock when the Central Powers abruptly ended the occupation of the Ukraine at the end of the Great War. Other planes came into the hands of the Whites through pilots who were seeking to join the anti-Soviet movement. In one example, in Austrian-occupied Odessa on September 16, 1918, a partial squadron under the leadership of Captain Evgenii V. Rudnev and, among others, Staff-Captain Vadim M. Nadezhdin flew from the Odessa aerodrome to Ekaterinodar (Krasnodar today) in the region of the Volunteer Army. In the fall of that year, Rudnev formed that army's Third Aviation Squadron, and Nadezhdin served as a pilot. On November 7, 1918, when Rudnev decided to return to Odessa, Nadezhdin assumed command of the detachment.[22]

Over time Nadezhdin's squadron became part of the Armed Forces of South Russia. It began with the Central Powers that occupied the Ukraine next door to the anti-Soviet Russians. At that point, it served the interests of the Central Powers to do everything they could to guarantee that the Red Army would not even think about, let alone plan, the retaking of occupied territory. To keep the Red Army focused on fighting the civil war, the Germans in particular occasionally provided White groups with military assistance. Naturally, the November end to the Great War terminated both the German occupation of the Ukraine and the transfer of war materiel to anti-Soviet Russians. When the British indicated that they would send large amounts of supplies to the Whites, they decided to lend their support to General Denikin, who headed the Volunteer Army. To gain a share of British largesse, the Don Cossack Army of General Vladimir I. Sidorin and the Caucasian Army of General Petr N. Vrangel' joined with Denikin over the winter of 1918–1919 to form the Armed Forces of South Russia. That meant that all three units would be beneficiaries of British military goods. Since General Denikin commanded the combined force, the shift elevated General Vladimir Z. Mai-Maevski to leadership over the Volunteer Army.[23]

Regardless of the new organization, British supplies did not start to arrive in southeast Russia until March 1919. The British did recognize that military aviation for Denikin's armies consisted of an unusual assortment of old German and Russian aircraft. Because White aircraft were worn out, the British tried to rectify such shortcomings by flying a squadron each of Sopwith Camels and D.H. 9s from Salonika (Thessaloniki), Greece, to

Ekaterinodar as a quick fix to strengthen the anti-Soviet air force. After March 1919, some D.H. 9s, more R.E. 8s, and a large number of Sopwith Camels arrived by ship. Built by the aviation firm founded in 1912 by Thomas O. M. Sopwith, the Camel was designed and tested in the first half of 1917 by Fred Sigrist and Herbert Smith. The "Camel" nickname for the F. 1 tractor biplane, based on the humped cover over the twin Vickers machine guns located in front of the pilot's cockpit, became official by common use. By 1919 the British considered the airplane to be surplus technology in peacetime, but there was nothing second-rate about the aircraft in the Russian Civil War. During the Camel's 16 months' military service in the Great War it racked up more combat victories than any other single type of aircraft.[24]

The RAF pilots who flew the 2 squadrons of Camels and D.H. 9s to Ekaterinodar stayed in south Russia and flew against the Red Army and Red Air Force. They were led by Commander Raymond Collishaw, an ace pilot who had destroyed 68 enemy aircraft during the Great War. Meanwhile, the Soviets had to respond to Denikin's armies and the improved aircraft situation for the Whites. The Red Air Force managed to assemble about 140 aircraft, spread among 23 detachments. Initially the Reds flew mainly SPADs and Nieuports or leftover German aircraft, such as the Fokker Dr I triplane. The Dr I was completed for construction in 1917 by Reinhold Platz on behalf of Dutch aircraft designer Anthony H. G. Fokker, who worked for the Germans. Interestingly, the head of the Red Air Force was Aleksei V. Pankrat'ev, a member of the nobility who chose to continue his aviation military career with the Soviets. In 1919 Pankrat'ev gave up his command over the Flying Ship Division for the Red Air Force leadership post. Far more common among Soviet pilots was the service of Aleksandr A. Chekhutov, who headed the Aviation Section on the Staff of the Red Army's Southern Front. His official military record confirms that he bore no relationship to the nobility.[25]

In the spring of 1919 the good news for the Whites was that the Red Army was badly overextended. It fought Kolchak in the east, Allied nations in the north, and various groups in the Ukraine. Unfortunately for the White cause the French Colonial Division in the southern portions of the Ukraine began withdrawing at Sevastopol' in April. In the southeast, the Tenth and Thirteenth Red armies, commanded respectively by Aleksandr I. Egorov and Ivan I. Kozhevnikov, failed to succeed in conducting offensives against the Whites. Denikin's armies then began moving northward and using aircraft—especially Camels—that could readily protect themselves in reconnaissance missions. In May Cossack troops and General Vrangel's cavalry defeated Red troops at Velikokniazheskaia. These forces then crossed the Volga River and severely mauled the Red Tenth Army; they also took thousands of prisoners. The victory opened the way toward the major city of Tsaritsyn (later to become Stalingrad and now Volgograd).[26]

In May the Whites had better aircraft than the Reds, incorporating advancements that they had made as military pilots, but their airplanes were out of date and extremely worn. Between April 4 and April 23, 1919, several British transport ships had arrived at the White-controlled Black Sea port of Novorossisk, carrying British aircraft that required rebuilding; in some cases wings had been detached from fuselages so that planes could be stored. Ivan P. Stepanov, an aviation inspector for the Volunteer Army and an expert at repairing planes, led a group of men in preparing the imported British aircraft for flight. Later in 1919 Stepanov served as aviation inspector for the Armed Forces of South Russia, as well as chief manager of airplane parts. Joining him in the inspector administration was Aleksandr A. Kovan'ko, who also was on the staff of the Military Aviation School, which supplemented experienced Russian Empire pilots with newly flight-trained men. Ivan N. Tunoshenskii headed the school. He had held a similar position in 1917 over the Aviation School of the All-Russian Aero Club, which also produced military pilots.[27]

As a result, when the White forces attacked the Tenth Red Army at the Cossack village of Velikokniazheskaia near the city of Rostov and the Don River they had air support from several squadrons, including the Fourth Don Cossack Aviation Squadron, commanded by Captain Fedor T. Zverev, and the First Kuban Aviation Squadron, headed by Viacheslav M. Tkachev. The leader of the Don Aircraft Division was Viacheslav G. Baranov, who had ended his pre–Bolshevik Revolution career as a lieutenant colonel and commander of the Seventh Russian Aviation Division. Before the end of the Russian Civil War, Baranov would be elevated to the rank of major general (*general-leitenant*). Initially the squadrons assisted ground forces by flying reconnaissance missions. Then, in the second week of May, two White squadrons of D.H. 9s, protected by RAF pilots in a squadron of Camels, conducted an air assault against the major Red Air Force aerodrome, which housed several aircraft detachments located at Urbabk.[28]

As suggested earlier, air battles were rare because of the limited number of aircraft and great distances across the several fronts. But as the White squadrons approached the Urbabk Aerodrome, a Red Air Force formation led by a Fokker triplane took off to meet and attack the White squadrons. The Red Air Force planes focused on the D.H. 9s, which had caused little damage to the aerodrome before the Red military aircraft fired upon the Whites. Two of the D.H. 9s went down in flames, but the Camels destroyed five Red fighters without any loss to themselves. Meanwhile, the Caucasus Army of General Vrangel' moved north to attack what was left of the Tenth Red Army at Tsaritsyn. A Red Army cavalry division of three thousand horsemen led by Semen M. Budennyi tried to relieve the city, but the British Forty-Seventh Squadron caught the horse-soldiers in a pass north of Tsaritsyn. The double machine guns on each plane slaughtered some of the cavalry. By the first of July 1919, the troops of Vrangel' had cleared the city of all Red Army soldiers. Several of Tsaritsyn's buildings suffered damage as a result of the conflict. Happily for the Whites, they inherited stockpiles of Red Army military supplies abandoned in the city.[29]

The success of Vrangel' and his Caucasus Army at Tsaritsyn prompted Denikin to celebrate the city's capture by issuing his "Moscow Directive" when he arrived on July 3, 1919. Vrangel' was ordered to take his victorious army north along the rail line west of the Volga River and turn west on the Vladimir–Moscow rail line at Nizhnii Novgorod. Sidorin's Don Army would follow the Voronezh–Riazan–Moscow line. Mai-Maevsky's Volunteer Army would march near the tracks of the Kursk–Orel–Tula–Moscow railway. As the Whites moved north in August and September, General Konstantin K. Mamontov and his Fourth Don Cavalry Corps slipped around behind the Red Southern Army Group and reached Tambov. To deal with the unexpected and unwelcome threat, the Red Air Force created a special aviation group of 25 aircraft under the leadership of Georgii A. Bratoliubov. It was an elite force of Moscow flight instructors and pilots such as Mikhail P.

Stroev, who had served the Russian Empire with distinction during the Great War. Its aircraft inventory included Il'ia Muromets aircraft that were the best flyable remnants of the Flying Ship Division. Iosif S. Bashko commanded the No. 1 Il'ia Muromets plane; No. 2 was flown by Vladimir A. Romanov, son of a cavalry captain and, despite the then-offensive surname, not a close relative of the last Romanov tsar.[30]

While the special aviation group caused casualties among the Fourth Don Cavalry Corps, surviving members moved south. But they did not endear themselves to local peasants and townspeople. In order to eat, drink, and also increase their personal wealth, cavalry members looted their way to Voronezh. On September 19 the cavalry broke through the Red Army line and returned to Denikin's forces. Although Mamontov's cavalry excursion behind enemy lines initially was viewed as an extraordinary victory, his horse-soldiers failed to do serious harm to the Eighth and Ninth Red armies, which were pressing hard against the defenses of Vrangel' at Tsaritsyn. Understandably, Vrangel' sharply criticized Mamontov for what he had failed to do militarily while his pillaging cavalrymen turned Russians against the Whites. Nevertheless, by the end of September, the Whites had safe ports behind them at Nikolaev and Novorossisk, and they controlled major cities of Kharkov, Kursk, Rostov, as well as Tsaritsyn. Denikin optimistically predicted that the Whites would be in Moscow before winter. Indeed, on October 14, 1919, the Kornilov Division (named for the deceased General Lavr G. Kornilov, who commanded White forces in 1918) occupied the city of Orel just 240 miles south of Moscow.[31]

Two days before the Kornilov Division captured Orel the White Northwestern Army had begun its advance to take the former capital city of Petrograd. The army was led by General Nikolai N. Iudenich, who had been the successful commander of the Caucasus Front against the Turks during the Great War. Iudenich never lost a battle in that conflict; only the Russian Revolution defeated him. Although hundreds of thousands of men had served under him between 1914 and 1917, his Northwestern Army was only the size of a single army division. It had its origins the previous year, when German troops occupied Estonia and allowed and even helped anti-Soviet groups. Estonia has a natural barrier border with Russia in the form of Lake Chud (today, Peipus). After the Great War closed, White forces moved east of the lake in a piece of territory along the lake's eastern shore. In 1919 and through Estonia, the British supplied a modest amount of martial hardware. White Russians tried to purchase military aircraft for Iudenich from the United States, but the American government had already sold or scrapped all surplus combat planes. The general's more serious problems centered on the fact that he only had 14,400 soldiers; they lacked discipline; many were Red Army defectors; and they marched with only 44 artillery guns.[32]

Nevertheless, on October 12, when this small army moved toward Petrograd, it enjoyed several favorable conditions. First, the Northwestern Army seemed to be synchronized with the planned advance of the Armed Forces of South Russia to Moscow. (Actually, Iudenich was aware of what was happening in the south, but there was no real coordination.) Second, the timing of Iudenich's march in the direction of Russia's former capital was a complete surprise to Petrograd's citizens. Third, the Red Army had recently transferred several well-equipped military units from Petrograd to the Southern Army Group to fight Denikin's forces, and they were not available to defend the city. Finally, the local Communist Party chief and chairman of the Petrograd Soviet, Grigorii E. Zinoviev, appeared to be confused and frozen over what to do. When the White soldiers occupied the palace towns of Pavlosk and Tsarskoe Selo on October 21 a forward unit could actually see the top of the golden dome of Saint Isaac Cathedral in the heart of Petrograd. At the same time the Northwestern Army made a serious mistake in not sending troops south of Petrograd to remove rails from the Moscow–Petrograd railway. Several days earlier War Commissar Trotskii had arrived by train and quickly reestablished Petrograd's defenses.[33]

Besides placing local civilians with rifles in defensive posts, the war commissar repositioned the Seventh Red Army in Petrograd and brought up the Fifteenth Red Army from below the former capital. Together the two armies totaled 73,000 men, giving the Reds a 5-to-1 advantage over the Whites. Moreover, the Reds possessed 13 times as many artillery guns. Trotskii also concentrated 87 aircraft in 17 detachments, which averaged 5 pilots for each squadron. Because Petrograd is on the coast of the Gulf of Finland, the Red Air Force planes included flying boats. Pilots carried out important battle tasks as part of their combined operations with ground troops. Reconnaissance flights tracked the extent of the White retreat each day. During daylight hours the Red Air Force conducted more than 300 combat missions during which aircraft-mounted machine guns wounded or killed White infantrymen. Red aircraft also dropped 2,500 kilograms (5,525 pounds) of bombs. Among the heroic aviation exploits on which the Soviet press focused was the flight of one of the flying boats, an old M-20 variant of the Grigorovich M-5. Pilot Karl Tekhtel' and his mechanic-observer, Aleksandr Bakhvalov, fired on White troops with their machine gun at a low altitude. Returning ground fire pierced the M-20's fuel tank, forcing the plane to land. Amazingly, the flying boat crew survived landing on soil rather than water. Whites captured and mistreated the two men. Reputedly before they were executed they shouted, "Long live Red Petrograd."[34]

The Seventh Red Army and the special collection of Red Air Force squadrons began their counterattack on October 21. Within three weeks the White Northwestern Army retreated to the Estonian border. That country permitted the army to enter, but only on condition that soldiers give up their weapons. In essence, the White army no longer existed. During 1920 the Soviet government negotiated and signed peace treaties with Estonia, Latvia, Lithuania, and Finland, formally recognizing the existence of these adjacent states, which had belonged to the Imperial Russian Empire. For Trotskii's role in defending Petrograd, the Soviet government named Gatchina after him and awarded him its highest decoration—the Order of the Red Banner. Meanwhile the Red Army, assisted by the Red Air Force, evicted the Kornilov Division from Orel on October 20. With more than 400,000 other troops, the Red Army forces actually in the southern field were 4 times larger than the remaining 98,000 White soldiers. And the Red Air Force flew more than 400 combat missions for several weeks after the start of the Red Army counteroffensive to harass White troops. The White armies retreated quickly. By the time the New Year opened in 1920, Soviet soldiers had captured Rostov and controlled the Don River.[35]

CHAPTER 10

SOVIET VICTORIES IN 1920 AND 1921

★ ★ ★ ★ ★

IN HIS AUTOBIOGRAPHY General Anton I. Denikin made no effort to hide the weaknesses that proved fatal to the White Army, known as the Armed Forces of South Russia. The Red Army had evolved into a massive, well-armed, and effective fighting force. By contrast, the White Army could not begin to replace soldier casualties or match the firepower of the Soviet military. Behind the White Army were large territories that it could not control. The government of South Russia never worked; Cossacks, for instance, often retained their own system of governance. Other major problems remained unsolved. Guerilla bands operated with impunity, anti-Jewish pogroms continued in full force, embezzlement and bribery flourished, and sundry peoples, such as those in the north Caucasus, refused to recognize the White administration.

The impact of these weaknesses was beginning to show outside the region as well. The dramatic retreat of the several White armies in the fall of 1919 had prompted the British to disband the two RAF squadrons that they had sent earlier. They soon would begin to shift their approach to Soviet Russia in commercial affairs. In June 1920 Leonid B. Krasin, about to be named Soviet commissar of foreign trade, was in London negotiating a commercial arrangement with the British government. That in turn led to the establishment of the All-Russian Cooperative Society (Arcos), Ltd., an Anglo-Soviet trade office in England, and served as background to the establishment of diplomatic relations. (Krasin later became Soviet ambassador to Great Britain.)[1]

The crews of the two RAF squadrons made their way to Rostov and in March 1920 were evacuated from the Black Sea port of Novorossisk, which had become key to the survival of a major portion of the White Army. On March 25–27, ships transported 34,000 White Army soldiers and some civilians to the Crimean Peninsula. As the Red Army approached Novorossisk, 22,000 more troops, along with many civilians who sympathized with the anti-Soviet Whites, were left behind; the lack of space on board departing ships made it impossible to evacuate them. Finally, those pilots who had endured combat missions and continued to serve flew still-functioning aircraft from the Novorossisk region to their new home in the Crimea. One of the most important of them was Viacheslav M. Tkachev. His performance as head of the First Kuban Aviation Squadron during the battles of Velikokniazheskaia and Tsaritsyn in May and June 1919 led to his promotion as a major general. Caucasus Army Commander Petr N. Vrangel' placed Tkachev in charge of the several aviation squadrons attached to his forces.[2]

It should come as no surprise that General Tkachev became the commander of White Army aviation in the Crimea. The peninsula had been preserved for the White movement by General Iakov Slashchev, who had kept remaining troops on the two connecting isthmuses in order to prevent Soviet soldiers from entering Crimea. Slashchev was fearless, most likely because he was a morphine addict, on the brink of insanity; he would later be relieved of command. Meanwhile, the Red Army leadership assumed that the White armies at Novorossisk had been beaten; the Reds failed to anticipate that many White Army troops would be evacuated. At the same time, most of the Red Army north of Crimea remained busy obliterating various guerrilla bands and anti-Soviet groups in the Ukraine. As it turned out, much of the Red Army soon had to ignore the peninsula. The Polish military began what would become a successful offensive toward the Western Dvina and Dnepr Rivers, and as a result the Red Army was forced to withdraw all available units and send them to deal with the Polish invasion; Moscow considered the Western Dvina and Dnepr rivers area to be Soviet Russian territories.[3]

The Polish offensive and the Red Army response brought changes to the White Army in Crimea. The retreat and mass transfer of the army prompted General Denikin to resign. At the beginning of April, General Mikhail Dragomirov led a war council in Sevastopol' at which members expressed a preference for General Petr N. Vrangel' as Denikin's replacement. Interestingly, and based on their personal histories, council officers did not want the selection of a replacement finalized by an election. On April 4, 1920, at Dragomirov's request, Denikin issued an order appointing Vrangel' as the commander in chief of the Armed Forces of South Russia. That same day Denikin boarded a British destroyer that carried him to Constantinople, where the White-controlled Russian Embassy had been converted into a hostel for Russian refugees. Denikin was reunited with his family, which eventually joined many other anti-Soviets in France. There the general wrote a multi-volume history of the White military. Denikin lived long enough to witness the German invasion of France in 1940 and the Allied liberation in 1944; he died in 1947.[4]

Another change resulting from the Polish-Soviet conflict was that for two months soldiers and pilots who had experienced long-term combat suddenly had peace and rest, giving those who were only mildly wounded time to recover from battle injuries. But the quiet and isolation of the peninsula also had its drawbacks. Only about forty older aircraft made the transfer from South Russia—a loss of 75 percent to 80 percent of airplane strength. Fuel, parts, and replacement aircraft from the British were limited or nonexistent. The Royal Navy did keep ships in the Black Sea that provided an escape route for Russian civilians and troops, but the British warned General Vrangel' in an ultimatum to "abandon the unequal struggle" with the Red Army. The British offered to negotiate with the Soviet government to win amnesty for White soldiers and the civilian population and provide evacuation and refuge for members of the White movement. But they told Vrangel' that if he rejected the ultimatum they would end all further military assistance.[5]

During this hiatus from war Vrangel' reorganized the 34,000 men who remained in the Armed Forces of South Russia. This vestige of the White

Army, renamed the Russian Army, was re-formed around one corps, a cavalry division, and a Caucasian brigade. At the same time, Tkachev created eight squadrons—a total that he reached only because Crimea still had several old Grigorovich flying boats and some Albatros planes that the Germans had left behind after the Great War. Senior experienced officers, one for each of these detachments, included Maksimilian E. Gartman, Anatolii T. Zakharov, Fedor T. Zverev, Aleksandr A. Kovan'ko, Vadim M. Nadezhdin, Aleksandr M. Pishvanov, Vladimir I. Strzhizhevskii, and Anatolii K. Timofeev. This fairly substantial air force belied the fact that on any given day fewer than half of the planes could actually take off and fly. Nevertheless, Tkachev worked with these aviation leaders to make sure that each squadron did its best to support ground troops. He also introduced a signal system to facilitate coordination between two or more aircraft during combat missions.[6]

Because the aviation component had to be self-sufficient, Tkachev established a supply administration to provide parts, tools, and fuel for aircraft, and an aviation park where damaged planes could be fixed and restored. As a result, a number of pilots ended up with tasks that kept them on the ground. Aleksandr A. Naumov, for example, spent his time building and testing bombs for aircraft; in between bomb-testing, he tested and approved airplane replacement parts. Sergei K. Modrakh, a pilot and military engineer, also worked in the supply administration, where he designed and created templates for aircraft repairs using locally produced materials. Ivan N. Tunoshenskii spent some of his time at the aviation park, where he taught officer-mechanics the secrets of how to give new life to old and heavily used aircraft engines. Despite serious efforts to overcome the shortcomings in Russian airplanes, Tkachev felt lucky any day he had two full squadrons of six aircraft each that could fly on a mission against the Red Army.[7]

Unlike the airmen, who could keep some airplanes flying, ground forces faced a more unsettling situation from the start. The reorganized White Army only had one hundred artillery pieces and about six hundred machine guns. The evacuation ships at Novorossisk did not have enough space for soldiers, civilians, horses, and heavier weapons. Worse, thousands of soldiers arrived in Crimea without rifles. Although the British kept ships in the Black Sea, they had no official plans to replenish ammunition, food, fuel, or military hardware. Fortunately, the Russian Army also had ships, several British tanks, armored cars, and four armored trains already stationed in Crimea or at the coast, and the trains could reach the mainland on the second, semi-artificial isthmus via a bridge. Meanwhile, the most immediate and very serious issue for Vrangel' was food. Crimea did not have food stored, nor could the peninsula raise enough of it to feed both the Russian Army and the civilian population. Just above Crimea were agricultural lands of the Northern Taurida region. The obstacle between the Perekop Isthmus and the Northern Taurida was the somewhat diminished and recently created Thirteenth Red Army—too small to invade the Crimea but big enough to stop a frontal assault by the Russian Army at the top of the isthmus.[8]

For the Russian troops and airmen, the hiatus ended at dawn on June 6, 1920. One corps of the Russian Army went north to the mainland in vessels and circled to the rear of the Thirteenth Red Army. A second corps took trains across the railway bridge east of the main isthmus to attack the enemy's flank. The third launched a frontal offensive from the Perekop Isthmus. As the battle raged Vrangel' told the British military mission that he was merely engaged with Soviet forces in order to secure food for his men. Yet, the British government was upset by what the Russian leader was doing. Indeed, Vrangel' had told the troops: "The Russian Army marches to liberate its native land from the Red vermin." Russian soldiers advanced employing their rifles, machine guns, artillery, tanks, armored cars, armored trains, and aircraft. The small Russian Air Force contributed to the well-designed onslaught by conducting reconnaissance missions, and added mayhem to the enemy's battle by spraying machine-gun fire and dropping bombs on Red Army troops. Three days after the start of the Russian offensive, the Reds retreated and Russian forces occupied Melitopol', the provincial capital of the Northern Taurida region.[9]

The Red Army high command tried to respond to the defeat of its forces, but the war with Poland placed severe limits on its military options. It sent a substantial Red Cavalry unit under the command of Dimitrii P. Zhloba, but Russian reconnaissance planes tracked the approach of Soviet horsemen to Melitopol' and reported it to ground troops. By coordinated attacks starting mid-June, Russian troops and aircraft virtually wiped out the cavalry shock group. The twin victories did have one negative impact: the British military mission in the Crimea withdrew. Nevertheless, the Vrangel' battle actions raised the confidence and spirit of his soldiers and secured food for the army. Just as important, the two defeats of the Reds also led to the acquisition of rifles, ammunition, 30 artillery guns, 2 armored trains, 3,000 horses, and 8,000 captive soldiers who "volunteered" to join the Russian Army. (Most had been conscripted, and the level of their loyalty was suspect.) The new Russian territory formed a curved arch around Melitopol' that ended at the Dnepr River in the west and at the town of Mariupol' by the Sea of Azov in the east. In 1920 some 3 million civilians resided in Crimea and Northern Taurida. Unfortunately for the White Army, the Red Army then was larger than the population that was under White control.[10]

The only reason that the Russian Army under General Vrangel' survived for almost seven months was that Poland invaded Soviet Russia. On April 25 the Polish Army began its advance across the western plains of the Ukraine. Within two weeks, Polish forces reached and occupied Kiev—a feat that Poland's King Boleslaw I had accomplished nine hundred years earlier. In one sense, the April 1920 invasion was simply an extension of Poland's 1919 takeover of Vilna (part of Lithuania today) and Minsk (part of Belarus today) as the Poles rejected the ethno-linguistic Curzon border. In the quest for Kiev, Marshal Jozef Pilsudski, chief of state and generalissimo of the Polish Army, coordinated his forces to force the retreat of the Twelfth and Fourteenth Red armies. Ironically, however, the Soviet troops retreated so fast that neither army suffered major losses. In fact, the apparently defeated Red forces were able to retain their organization, troops, and weapons, and would be used for a later counterattack against the Poles in Kiev.[11]

The Polish Army's march to Kiev required aircraft, primarily for reconnaissance. In November 1918, as the Polish Army disarmed German soldiers and occupied the Warsaw capital of the newly revived country, it also took over the Mokotov Aerodrome and the remaining German aircraft. The Germans had used the aerodrome as a flight-training center, which also contained workshops to repair and service airplanes. With the closure of the

Warsaw branch of Albatros-Werke, Polish aircraft workers manned the aircraft workshops at Mokotov, forming the basis for the creation of the Centralne Warsztaty Lotnicze (Central Aviation Workshops). The facility, which began operations on January 1, 1919, was subordinate to the Aerial Navigation Section of Poland's Ministry of Military Affairs, and was headed by Karol Slowik, who oversaw 300 Polish workers. The group's first major effort focused on producing the Hannover-Roland CL IIa aircraft that Germany's Hannoversche AG had subcontracted to Luftfahrzeug Gesellschaft (Roland). The air tractor-powered biplane was equipped with a 220-horsepower Austro-Daimler engine. Slowik received documentation on the plane from the subcontractor and prepared specifications for templates of the airframe.[12]

A terrible mistake occurred during the building of the Hannover-Roland CL IIa. No structure or stress tests were undertaken. Later in 1919 a test pilot, Lieutenant Kazimierz Jesionowski, flew the biplane successfully several times. During an official demonstration flight for Marshal Pilsudski and army officials, the aircraft—flown by Jesionowski, with Slowik in the observer cockpit—disintegrated during an aerobatic display, killing both men. From then on, the Centralne Warsztaty Lotnicze focused on overhauling abandoned or otherwise secured German aircraft, including some that had been damaged. By mid-1920 the workshops employed about 500 people, who repaired or rebuilt 544 airframes and an equally large number of airplane engines and wooden propellers. Many of the renewed aircraft were Albatros fighters. Thus when the Red Army counterattacked the Polish Army actually had more planes than the Red Air Force.[13]

The Twelfth and Fourteenth Red armies, which were closest to the Polish troops occupying Kiev, belonged to the Red Southwestern Army Group, commanded by Aleksandr I. Egorov, a former metalworker. One of the group's units was the First Red Cavalry Army. In Soviet military circles, the unit was commonly referred to as the Konarmiia ("Kon" comes from the Russian word *konnitsa*, or cavalry). Semen M. Budennyi led the Konarmiia, which had six divisions and three aircraft squadrons; Lev K. Vologodtsev was commander of the cavalry air fleet, and Leonid A. Kul'tin was the assistant chief of staff. Battling the Poles at the end of May, the cavalry riders had to dismount and fight on foot, like the infantry. By June 12 the Konarmiia had broken through Polish lines north and south of Kiev, forcing the Polish Army to abandon the city to Red Army control. As the Poles retreated, the Twelfth and Fourteenth Red armies followed the Konarmiia westward. In one sense a more important offensive came from the Red Western Army Group, which began its drive against the Poles from White Russia. Mikhail N. Tukhachevskii took charge of the group, which comprised the Third Cavalry Corps and the Third, Fourth, Fifteenth, and Sixteenth Red armies.[14]

As the Red Southwestern Army and the Red Western Army pursued the Polish Army, the two Soviet groups had assistance from the Red Air Force, which included 210 aircraft apportioned among 51 understaffed detachments. On the plus side, more than 70 percent of the Soviet Russian planes were recently manufactured imports or (some) copies of the Sopwith 1 ½ Strutter. The detachments were formed around well-prepared pilots, who often flew similar types of aircraft, powered by engines that were fueled with gasoline, which became available as fuel for aircraft after the White lost the Caucusus. Special aviation squadrons had been established for the Konarmiia and the Western Army, among them the Saltanov, Mogilëv, and Mozhaysk detachments, which together involved thirteen squadrons. Red Air Force missions included reconnaissance and photographs of Polish defensive positions, machine-gun attacks, and the dropping of 6,400 kilograms' worth (14,800 pounds) of bombs on Polish troops. Soviet Russians claimed that by attacking Polish aerodromes at Orekhov, Zhodin, Post-Volynsk, and others, the Red Air Force reduced the effectiveness of enemy aircraft and provided a screen against them. At the same time a few Red Air Force pilots, such as Andrei D. Shirinkin and Georgii S. Sapozhnikov, shot down several Polish aircraft. Altogether, the Red Air Force conducted 2,100 combat missions against Polish ground and aviation forces.[15]

By the second week of August, Tukhachevskii's Red Western Army Group had reached the Vistula River near Warsaw. Marshal Pilsudski reorganized a number of Polish divisions and, with help from scouting aircraft, knew the position and extent of the Red group. In a brilliant move on August 16, Pilsudski sent divisions flanking and threatening to surround the opponent. Tukhachevskii's forces retreated, but suffered tremendous losses. Other Polish divisions outflanked the Red Southwestern Army

Group by the Niemen River. The Polish victories resulted in an armistice and peace talks that eventually took place in neutral Riga, Latvia, beginning on September 21, 1920. The negotiations, which continued for several months, produced the Treaty of Riga on March 18, 1921. Neither the Poles nor the Soviet Russians secured the borders they wanted, but Poland did end up with territory beyond the Curzon Line, which it kept until the Nazi invasion of 1939.

During the Polish-Soviet war of 1920 France supported Poland, but it also backed the Russian Army under Vrangel'. Political leaders in Paris hoped to contain Soviet Russia and prevent the expansion of communism. Unlike the British, the French promised assistance and officially recognized the de facto Russian government in the Crimea.[16]

The reality proved quite different, however. French prime minister Alexandre Millerand awarded de facto recognition on August 10, 1920, but with conditions that included repayment of Russia's huge war debt to France: the aid would only come with the extension of credits based on ironclad security from the Russian government in Crimea; in short, Vrangel' could not rely on receiving any aid from France. Fortunately, the harvest of "corn" (wheat) in Northern Taurida produced abundant reserves that would feed both civilians and troops in the occupied area. Indeed, a surplus allowed for the exchange of wheat for ammunition, clothing, coal, gasoline, and oil. The purchased items came from Bulgaria and Romania and were transported in Crimean ships. Gasoline was critical for a number of Russian Army vehicles, of course, including airplanes. Happily for the Whites, the British had supplied quite a few D.H. 9s, and the Russian Air Force retained them even after British pilots had withdrawn from the Armed Forces of South Russia. Unhappily, however, these heavily used aircraft would never be replaced. Their service put a premium on repairs and forced the cannibalization of worn-out planes for parts, engines, and propellers.[17]

Meanwhile, the Soviet command concentrated on forty-five aircraft that formed two groups to help gather intelligence on the Russian army that occupied and defended the Northern Taurida. The small number of Red Air Force aircraft reputedly consisted of seventeen squadrons, but there were only two to four planes and pilots in each detachment. The bulk of the Red Air Force, with its well-trained pilots, focused on the Polish enemy. Even more recent Soviet commentary on this early era admits that the Russian Air Force of 1920 had better aircraft than the Soviets, who saved their best planes for the conflict with Poland. Regardless, in July near Melitopol' an unusual event—an air battle between opponents—occurred. Remarkably, the battle involved the two top leaders of the opposing forces—Viacheslav Tkachev, leader of the Russian Air Force, and Petr Mesheraup, head of the Red Air Force group above the Northern Taurida. Tkachev, flying a D.H. 9, and Mesheraup, in a Nieuport, exchanged machine-gun fire and conducted evasive or attack flights for an extended period. Both planes took hits from bullets, but neither the pilots nor their aircraft suffered a decisive blow. After three-quarters of an hour, the two pilots broke off combat, and each headed for his respective home base.[18]

When the duel between opposing pilots and aircraft took place, the Russian Army had secured the Northern Taurida. As already indicated, the victory had enabled the Russians to replenish their military hardware, solve the food shortage problem, and supply some (uncertain) replacements for the significant casualties they suffered in battle. No matter how the Polish-Soviet war ended, the Russian Army had to be concerned about how to expand the number of its troops with loyal and committed soldiers who would be able to face an eventually enlarged Red Army. Vrangel' actually considered withdrawing from the Northern Taurida, keeping a smaller Russian Army unit to defend the Crimea at the eighteen-mile-long Perekop Isthmus, and returning the rest of the army to the Kuban region between the towns of Novorossisk and Yeist to the north. Vrangel' hoped to make Ekaterinodar the headquarters for cleansing the region of the small but numerous Red Army detachments. But most important, the general knew that the Kuban could provide food, supply many Cossack horsemen, and enable the White Russians to resume the struggle against the Soviets—this time from Cossack lands north of the Caucasus.[19]

On August 13, 1920, approximately 5,000 Russian Army troops, mainly Cossacks under General Sergei G. Ulagai, embarked simultaneously from the Crimean ports of Kerch and Feodosiia. The men and their equipment went ashore at the two Kuban towns of Primorsko-Akhtarskaia and Anapa. An important aspect of the operation was to join and merge with local

Cossacks, who despised the Soviets and the Red Army. For this apparently preliminary invasion White soldiers were aided by 130 machine guns, 28 pieces of artillery, several armored cars, and a large squadron of 8 aircraft. Aleksandr A. Kovan'ko, who became the commander of the First Aviation Squadron on June 29, led the detachment that provided cover and protection for Ulagai's forces. The squadron conducted 12 reconnaissance flights and 3 group bombing missions against the Red Army. Initially, aircraft also flew messages back and forth between the Kuban and Crimea.[20]

The key object of the overall operation in the Kuban territory was the seizure of Ekaterinodar. With that in mind, General Ulagai quickly departed from Primorsko-Akhtarskaia and left behind substantial provisions and ammunition. The Cossack-dominated Russian Army went inland and reached the Timoshevskaia railroad junction, about thirty miles north of Ekaterinodar. Delaying in order to cement contacts with local Cossacks and organizing troops, Ulagai spent several days at Timoshevskaia, angering Vrangel', who was left waiting at the Crimean port of Kerch and sent aircraft to demand detailed information on Ulagai's situation and plans. As might be expected, those few days at Timoshevskaia enabled several Red Army detachments to merge, confront, and battle the Russian Army. In the process Ulagai lost the initiative, retreated to Primorsko-Akhtarskaia, and re-embarked on ships to Crimea in the first week of September. The failure of the expedition crushed the Vrangel' plan for using the Kuban lands as the center of Russian Army activity. Despite the loss of several hundred men as well as horses in battles with the Soviets, the Russian Army actually overcame its losses—by bringing 1,500 recruited Cossacks and 600 horses back to Crimea.[21]

At virtually the same time as the failed expedition into the Kuban region, Poland delivered its successful counteroffensive against the Red Army. The Polish victory was a serious blow. As noted earlier, the Polish-Soviet conflict had ended in armistice in September, and negotiations later had determined the border between the two countries. On paper that meant that the multi-million-man Red Army was now free to bolster its military units around the periphery of Northern Taurida. Vrangel' knew that for the Russian Army to survive he soon would need outside assistance. To prove the seriousness of the Russian Army—as well as its desperate need for materiel—he took military representatives from France, Great Britain, Japan, Poland, Serbia (Yugoslavia), and the United States, along with several foreign newspaper correspondents, on a railroad and automobile tour of military installations in the Northern Taurida. (In his autobiography, he incorrectly listed all the officers as members of official military missions.) Vrangel' could show the men sophisticated and elaborate fortifications built by the Russian Army without technical equipment. At the Akimovka Station, just south-southwest of Melitopol', the foreign representatives left the train to inspect reserve troops and visit an aviation squadron at an aerodrome.[22]

The squadron's commander was General Tkachev, who also headed the much-reduced Russian Army Air Force. In September the squadron focused primarily on conducting reconnaissance missions. On sunny days, aircraft would fly across enemy lines but maintain a high-enough altitude to avoid damage to planes or injury to pilots by small-arms fire from the ground; pilots or observers could see and report on the location or expansion of Red Army forces. During the visitation by foreign military observers, Tkachev took off on a demonstration flight during which he conducted a number of well-executed maneuvers. Subsequently, Vrangel' commented that the controlled movements "were all the more remarkable because most of the machines [aircraft] were in a very bad state of repairs, and only the incomparable boldness of the Russian officer made up for the deficiencies of his material." After the flight, the military men crowded around Tkachev and expressed their admiration for the piloting skills that he had displayed.[23]

In turn, Tkachev spoke honestly when he admitted that his Air Force consisted mainly of worn-out planes. In essence, he made a plea for new aircraft from the countries represented by the military officers; otherwise, the Russian Army Air Force would soon become non-functioning. Vrangel' supported his Air Force commander's comments and pointed out that he already had made efforts to secure aircraft elsewhere. For example, he had sent pilot Maksimilian Gartman, who had recently been promoted to aviation administration, to Bulgaria to purchase aviation equipment—including a number of planes that Bulgaria had received from Germany. Regrettably, however, the aircraft that he wanted had been destroyed by mistake—through orders from the Allied Commission of Control that operated in Bulgaria at the end of the Great War. The commission members who were responsible for the order were men from England. Although the military officers who participated in the tour were very impressed with the Russian Army and the Air Force, their positive views had little impact on their governments. Fortunately, the Ministry of Commerce and Industry in Crimea bought 10 million pouds (a *poud* is about 40 pounds) of Russian grain. The export of 2.5 million pouds of grain and the money it generated enabled the Russians to buy enough coal and oil to keep in reserve to power evacuation ships if needed.[24]

As the Russian Army Air Force spiraled down, the Red Air Force above the Northern Taurida increased both the number and the quality of its pilots and planes. At the end of 1919 Soviet army and aviation leaders met and issued a directive calling for an end to all token aviation operations that proved to be ineffective. In short, selected important fronts would receive a concentration of military air resources large enough to enable them to achieve air superiority and to apply deadly harm to enemy ground troops through reconnaissance and combat attacks using bombs and machine guns. The application of the directive guaranteed that the Red Air Force would be much stronger than the Russian Army Air Force. And Red pilots were well-trained. Soviet sources point to pilot Nikolai N. Vasil'chenko as an example. At one point in August he reputedly conducted an air battle with a squadron of seven White aircraft. He brought down one plane and compelled the other six to turn around and go home to their aerodrome. During the late summer and early fall the Red Air Force and Red Air Fleet added enough squadrons to give Soviet aircraft overwhelming air superiority.[25]

Among the units that were added between late summer and early fall was the Fourteenth Aviation Squadron. It included a special pilot-observer, Garal'd A. Matson, a Latvian from Riga. His lower-class, petty bourgeois status did not prevent him from enjoying success as an officer during the Great War. He graduated as a second lieutenant from the Military Topographical School, completed the photogrammatical course in Kiev, attended the Military School for Pilot-Observers, and attained the rank of staff-captain by 1917. Matson's training made him an expert at photographing enemy forces and then creating a precise and detailed topographical map of the opposition. His work guided Red Army units effectively both in deciding where to attack the Russian Army and in firing their artillery.

The Red Army's battle to take the Crimea also marked the first time that the Red Air Fleet had played a prominent role in a major attack. Before this—during the Russian Civil War, for example—navy aircraft units, composed mainly of Grigorovich flying boats, had been confined to limited missions on Russian rivers that had had only a small impact on the various battle fronts.[26] By contrast, in October 1920, from Taganrog and Mariupol' in the east, and Odessa and Nikolaev in the west, the flying boats took off several times a day to complete reconnaissance missions; they also dropped light bombs on Russian Army fortifications and fired machine guns on troops and cavalry formations. Experienced sea pilots such as Nikolai S. Mel'nikov, Evgenii I. Petkevick, and Vasilii G. Chukhovskii had little to fear from the dwindling number of patched-together planes in the inventory of the Russian Army Air Force. Adding insult to injury, the Red Air Force also brought in the last flyable Il'ia Muromets. Piloted by Aleksandr K. Tumanevskii, the large aircraft carried a crew of soldiers who manned machine guns. Although the Il'ia Muromets apparently did not shoot down any White airplanes, Soviets claimed that the multi-engine craft managed to damage at least four aircraft belonging to the Russian Army Air Force. During the Soviet conquest of Crimea, the Red Air Force and Red Air Fleet participated in more than a 1,000 missions and dropped more than 1,600 kilograms' worth (3,520 pounds) of bombs.[27]

Before the end of October, Mikhail V. Frunze, who served as commander, collected and deployed the Fourth, Sixth, and Thirteenth Red armies and the First and Second Red Cavalry armies. Troops and cavalrymen were stretched above and all across Northern Taurida, from the west at Kherson at the mouth of the Dnepr River to the east at Nogaisk on the Sea of Azov. The massive presence forced the Russian Army to give way as White soldiers moved south to the Crimea. By the end of the first week in November, the Red Army of 188,771 men attacked the 26,000 White soldiers and 16,000 poorly armed reserve troops at the Turkish Wall defenses at the Perekop Isthmus. Since the Russian Army temporarily stopped the Reds at the main isthmus, nature awarded the Red Army victory. Unusually cold weather froze both the water and the mud in the four-mile stretch called the Sivash that led to Crimea's Lithuanian Peninsula. The resulting fog enabled the Red Fifteenth and Fifty-Second Rifle Divisions, along with the 153rd Rifle and Cavalry Brigade, to cross unseen into the peninsula.[28]

The success of these Red Army divisions in taking and occupying the Lithuanian Peninsula resulted in a threat to the rear of the Russian Army, forcing it to retreat from the Turkish Wall and move to another line of fortifications farther south. Unlike the Vrangel' performance in the chaotic and incomplete departure of troops and civilians from Novorossisk in March, he had carefully planned an evacuation of the Crimea should the Turkish Wall fall to the Red Army. On November 11 he ordered White ships to go to embarkation points at Evpatoriia, Sevastopol', Ialta, Feodosiia, and Kerch. And he asked the ships from France, Great Britain, and the United States—both at Constantinople and in the Black Sea—to assist with the evacuation, which took place from November 14 to November 16. Departures began in Sevastopol' and ended in Kerch. Some 126 ships carried 145,693 military personnel, including wounded soldiers, as well as civilian men, women, and children to Constantinople. Among a sample of surviving pilots, Fedor T. Zverev ended up in a Cossack village in Lebanon; Viacheslav G. Baranov spent time in Bulgaria, Serbia (Yugoslavia), and France before moving to England in 1940; Evgenii V. Rudnev died in Paris in 1945; Aleksandr A. Naumov and Maksimilian E. Gartman joined the Legion of Former Russian Pilots in the United States; and Aleksandr A. Kovan'ko and Anatolii T. Zakharov, and Viacheslav M. Tkachev went to Yugoslavia. After World War II Tkachev returned to Soviet Russia, where he was arrested and sentenced to prison. Released in 1955, he returned to the region where he had grown up, working at an industrial cooperative in Krasnodar (formerly Ekaterinodar), where he was a bookbinder. He died in the Soviet Union ten years later.[29]

After the evacuation of the Russian Army and its pilots from Crimea, only a modest number of Red aircraft participated in retaking Central Asia and Siberia east of Lake Baikal. The last important event in the Russian Civil War occurred in 1921, with the uprising at the Kronshtadt Fortress on Kotlin Island off the coast from Petrograd. Initially, Kronshtadt sailors had been the strongest allies of the Bolshevik-Communist Revolution. But in the winter of 1921 they turned against the Soviet government by supporting peasants and workers, especially in the worker strikes in Petrograd. Sailors denounced Soviet leaders for betraying the revolution, terrorizing opponents, and stealing the peasants' food. From March 7 to March 18, they fought the Red Army directed by Trotskii and led by Tukhachevskii. Two divisions crossed the ice-covered Gulf of Finland toward the fortress. Sailors responded by sending 2 aircraft equipped with machine guns and by firing large naval guns; the latter caused serious casualties among Soviet troops. The Red Army responded with artillery, and the Red Air Force carried out 137 weapons-loaded aircraft missions. The costly Red victory had unfortunate consequences for many of the 14,000 sailors; 45 were executed and many others were imprisoned. This final episode of strife occurred in the midst of the Tenth Communist Party Congress. The sailor revolt that accompanied worker strikes and peasant rebellions prompted Lenin to recommend drastic measures to restore the exhausted economy. Congress approved what was soon called the New Economic Policy. It permitted the limited return of private capitalistic enterprise as it terminated the requisitioning of peasant food.[30]

CHAPTER 11

AIRCRAFT DEVELOPMENT, 1918–1924

☆ ☆ ☆ ☆ ☆

AT FIRST BLUSH it may seem that the Red Air Force conducted an impressive array of battle operations during the Russian Civil War. Soviet records show, for example, that it flew 21,421 missions and dropped up to 94,508 kilograms (208,863 pounds) of bombs. Yet, to put things in perspective, these totals fell well below the comparable statistics recorded by the military aircraft of France for the single month of April 1918, during the Great War. Moreover, unusual for the record of any country, some 9,000 aircraft of the Red Air Force missions focused primarily on distributing propaganda to promote the work, the views, and even the presence of the Soviet government—an effort that proved effective in peripheral areas, especially North Russia. At the conclusion of the civil strife, the Red Air Force had 325 aircraft in 54 detachments, with an average of 6 aircraft and pilots for each squadron. The squadrons were supplemented by 13 units of various numbers of seaplanes belonging to the Red Air Fleet. Together, they constituted a formidable air component. More significant for the future military aviation in Soviet Russia, however, was the government's serious endeavor to standardize equipment and enhance the technological advancement of aircraft.[1]

As noted earlier, one of the most important institutions for the technical foundation of the Red Air Force was what ultimately became the Central Aero-Hydrodynamic Institute, established on December 1, 1918, by Nikolai Egorovich Zhukovskii, with the backing and encouragement of Lenin. Known best by its Russian initials, TsAGI, the Institute clinched Zhukovskii's reputation as the father of Russian aviation. Biographical information about him is provided in chapter 1, but it is appropriate to mention here that he established an aerodynamics laboratory and published an important work on the principles of flight. During the Great War he put together a classroom course from that text for would-be military pilots who studied aeronautics at Zhukovskii's Moscow School of Theoretical Aviation. After the conflict, in 1919, he founded the Red Air Force Engineering Institute, which was renamed the Zhukovskii Military Air Academy after his death in 1921.[2]

Centered initially on Voznesenskaia Street in Moscow, TsAGI operated with seven departments—theoretical, aviation, engines, communication, design, aero-hydrodynamics, and research. The Soviet Supreme Council of National Economy awarded the new organization a substantial budget used to pay personnel, purchase equipment, and publish a journal that focused on its accomplishments. The first issue of *TsAGI Works* came out on December 10, 1919. Early projects included the creation of an aerodynamic laboratory and the construction of a huge wind tunnel to test how aircraft would perform in the atmosphere. One of the first major airplane concepts that attracted Zhukovskii's attention and that of his former student, Andrei Nikolaevich Tupolev, concentrated on a new multi-engine replacement for the Il'ia Muromets, the reconnaissance-bomber that was so successful in both the Great War and the Russian Civil War. Even the last flights of the surviving Il'ia Muromets aircraft had made a contribution to the defeat of White forces in the Crimea. After that, TsAGI put priority on building a large plane to serve as a substitute reconnaissance-bomber in the Red Air Force inventory.[3]

Zhukovskii headed the Special Committee for Heavy Aviation and, together with Tupolev, engineered the proposed new aircraft—a TsAGI-built, twin-engine triplane—with grandiose plans to build 30 copies. When the prototype was completed in 1922, the design proved to be a failure.

Zhukovskii had died by then, and Sergei Alekseevich Chaplygin, a mechanical engineer and graduate of Moscow University, had filled his slot. Tupolev continued his own work at TsAGI by designing and building in 1922 a small single-engine monoplane that included metal (duralumin alloy) parts. Powered by an Anzani motor, the new ANT-1 flew at a maximum speed of only 135 kmh (84 mph), with a range of 450 km (280 miles)—a relatively poor performance even for 1922. Within 2 years, however, Tupolev had drafted a twin-engine, all-metal mono-wing bomber—the first in the world. This one, the ANT-4, fulfilled the objective of the Special Committee for Heavy Aviation, and became the basis for 38 large aircraft that Tupolev designed before the start of World War II. Large aircraft became Tupolev's life-long specialty.[4]

Like Tupolev, virtually every important aircraft designer or engineer in Soviet Russia spent time at and acquired knowledge from TsAGI. The Institute served as the focal point for the development of aviation technology. At first, however, the serious obstacle to producing decent military aircraft did not come from any dearth of technology. The real problem was the lack of facilities for building airplanes. Joseph Stalin, a senior political commissar of the Red Army and an original member of the powerful Political Bureau (Politburo) of the Central Committee of the Communist Party, addressed this shortcoming in a message to the Politburo on August 25, 1920. Before the end of the Russian Civil War, Stalin urged the Politburo to take whatever action was necessary to create a viable aviation industry that could build well-performing military aircraft. His initiative, supported by the Politburo, prompted the Soviet of Labor and Defense to issue a decree that secured experienced workers for a resurrected aviation industry. All men under the age of fifty who had worked at least six months in an aircraft factory had to return to their previous work stations.[5]

An important benchmark in the push for the creation of Soviet aircraft factories came on June 22, 1920, with the formation of the Commissariat of Foreign Trade. An Anglo-Soviet trade office—the All-Russian Cooperative Society, Ltd. (Arcos)—already had been established in London. Within a month, a branch of Arcos opened in the United States and was soon joined by a subsidiary of a Russo-British barter agency called the Products Exchange Company, or Prodexo. The new Soviet trade organizations soon began selling Russian platinum, used in jewelry and electrical connections, as well as wood, which was commonly employed in manufacturing paper products, including cardboard boxes. Such sales provided hard currencies—both U.S. dollars and British pounds—that could be used in America or England as well as most other countries. The Russian ruble was (and still is) a soft currency that could only be used inside Russia. In the 1920s, the ruble was generally pegged at a rate of five rubles to one U.S. dollar. The exchange of Soviet products for foreign goods through Prodexo and the acquisition of hard currencies through the sale of Soviet goods and commodities via Arcos enabled Soviet Russia to import special timber, metal products, aircraft linen, fuels, machine tools, and various other equipment used in the manufacture of airplanes in the early 1920s.[6]

All of these efforts to promote foreign trade and reacquire experienced aviation workers were challenged by a shortage of experts and raw materials, and by a series of worker strikes in 1921. Only gradually would the New Economic Policy of 1921 allow for an economic recovery needed to establish a state-run system of aviation factories and workshops. The formation of the Chief Directorate of Nationalized Aircraft Factories (*Glavkoavia*) in 1920 failed to overcome these various problems. But in 1921 the Soviet government chose to spend several million gold rubles to resurrect Russia's aviation industry. The investment of government funds especially helped State Aviation Factory No. 1, or GAZ-1 (the former Dukh Company in Moscow), which was headed by chief engineer Nikolai Nikolaevich Polikarpov. It was only the aircraft-building facility that continued to function after the revolutionary year of 1917. At first, the factory's main effort focused on reproducing as best it could the Sopwith 1 ½ Strutter.[7]

Polikarpov was born in 1892 in a town east of Saratov. After completing the equivalent of high school (*gimnaziia*), he entered the Saint Petersburg Polytechnic Institute, graduating as a mechanical and aero engineer in 1916. The Petrograd branch of the Russo-Baltic [railroad] Wagon Company hired him immediately; the factory built aircraft designed by Igor Sikorsky. Polikarpov applied his engineering skills to one of the later Great War versions of the Il'ia Muromets, but he spent more of his time contributing aero-engineering refinements to the building of the Sikorsky S-20 and the licensed construction of the Nieuport 17 fighter aircraft. A single-engine biplane, the S-20 is interesting because it incorporated characteristic features of the S-16, such as the fin and rudder, as well as the top wing of the Nieuport 17. As a fighter plane, the S-20 had a Vickers machine gun, mounted in front of the pilot on the engine cowling, that fired synchronically through the rotating propeller. The aircraft provided escort duties for Il'ia Muromets reconnaissance-bombers. In 1918 the revolution and economic collapse terminated aviation production at the Russo-Baltic plant in Petrograd.[8]

Polikarpov joined several workers from the non-functioning Lebedev factory as the men traveled to Moscow. At this point in 1918, the soon-to-be-renamed GAZ-1 sought both workers and materials to build the Sopwith aircraft. Polikarpov, however, focused much of his attention on the D.H. 4, which the British had licensed to Dukh shortly before the Russian Revolution. He then spent time examining carefully a D.H. 9 that had been captured from North Russia. Based on his inspections, he made changes to the original D.H. 4 that led to the building of the first 20 R-1 (*razvedchik*—reconnaissance type-1) aircraft in 1920. The initial version of the R-1 used different engines, such as the 240-hp Fiat motor. Over time, Polikarpov made numerous improvements in the military plane, including the installation of the M-5 Russian motor, which copied the U.S.-made Liberty engine. With this powerful engine, the R-1 could carry 2 men, 3 machine guns, and up to 400 kilograms (884 pounds) of bombs. During the decade of the 1920s, state aviation factories produced 2,447 R-1s, including floatplanes built at GAZ-10 in Taganrog.[9]

Meantime, the Soviet government formed a special commission that in 1921 invited airplane designers to submit proposals for an indigenous fighter (*istrebitel'*) aircraft for the Red Air Force. The invitation drew two proposals—one from Mikhail M. Shishmarev, and the other from Dimitrii P. Grigorovich; neither was accepted. Even so, Grigorovich had a strong and early connection with TsAGI, and by 1921 he had joined Moscow's GAZ-1 as its technical director. Although his Istrebitel' No. 1 failed to win a contract, he spent the next three years working to improve and redesign the plane. With the help of workers at GAZ-1, Grigorovich constructed what became Istrebitel' No. 2, or I-2. It was a smooth-looking, aerodynamic biplane that embodied major alterations, including a welded steel-tube frame that braced the fuselage, around which a secondary wooden structure was built.[10]

The I-2 carried two PV-1 (Pulemet Vozdushnii No. 1) machine guns, placed on top of the fuselage, forward and in front of the cockpit, in order to make breeches accessible to the pilot. The machine guns imitated the British Maxim- and Vickers-type weapons that fired 7.62-mm (.300 inch) caliber ammunition. The I-2 had an M-5 motor that represented a modified U.S. Liberty engine, which provided it with a maximum speed of 235 kmh (146 mph) and a range of 650 km (400 miles). Pilot flight tests had been conducted by Andrei I. Zhkov. GAZ-1 built 164 of the planes; GAZ-3 constructed an additional 47 of the I-2s in Petrograd (which became Leningrad in 1924, after Lenin's death and the creation of the new Union of Soviet Socialist Republics (USSR), which had been established by a ratified constitution that year).[11] Before Grigorovich died, in 1938, he designed dozens of other aircraft, from flying boats and sleek fighters to a bomber that for its time (1931) was the largest plane built in the USSR.

In reality, Istrebitel' No. 1 did not come from Grigorovich, but from the work of his colleague, Polikarpov, who in turn had help from technical specialist Ivan M. Kostkin and from Aleksandr A. Popov, the head of manufacturing at GAZ-1. The prototype for the first Soviet fighter to enter production received the designation Il-400—the "I," obviously, for Istrebitel' and the "-400" for the 400-hp Liberty engine that powered the aircraft. A low monoplane craft, the Il-400 had a modern appearance, especially when compared with a typical biplane of the period. As might be expected, it bore some resemblance to the R-1 aircraft without the top wing. The aircraft's first flight took place on August 14, 1923, with Konstantin K. Artseulov as the test pilot. But the actual flight time was very brief. Despite Artseulov's best efforts with the control stick, the aircraft shot up into a stall right after takeoff. The plane fell like a rock, but fortunately Artseulov survived the crash. Subsequent wind tunnel tests, which should have been made before the failed flight, revealed that the center of gravity was too far back on the fuselage. Polikarpov redesigned and improved the aircraft by pushing forward the engine and cockpit and as adding some metal to the previously all wood and fabric version. After test flights in 1924, the Red Air Force contracted to secure 33 of the Il-400B. The aircraft remained difficult to fly, but

Polikarpov became famous in Russian aviation circles for his numerous designs of fighter and twin-engine military aircraft.[12]

The fourth major aircraft designer who gained prominence for his aviation achievements was Sergei V. Il'yushin. Born in a peasant family near Vologda in 1894, he developed into an aircraft mechanic at the Komendantskii Aerodrome in Saint Petersburg. He applied his technical skills in servicing the Il'ia Muromets reconnaissance-bomber during the Great War. Drafted into the Imperial Russian Army in 1916, Il'yushin was immediately spotted as a man with intimate knowledge of airplanes. As a result, the young soldier was assigned to a flight-training unit, where he completed ground school and finished the course as a pilot in 1917. An enthusiastic revolutionary, he continued service in the Soviet army as an aviation mechanic, riding a train in a car dedicated to repairing aircraft for the Red Air Force. Il'yushin ended up on a train supplying and assisting Red Army and Red Air Force personnel in North Russia. There, when a British Avro 504K was captured by the Red Army after the pilot was forced to land on Soviet-held territory, Il'yushin was given responsibility for recovering the plane and delivering it to GAZ-1.[13]

At GAZ-1 the plane was dismantled and Soviet technicians took precise measurements and made drawings of the Avro 504 K to use in crafting templates so they could construct copies of the aircraft (as *Uchebnyi samolet* No. 1) to produce the Soviet U-1 training plane. Between 1923 and 1931, GAZ-5 in Moscow and GAZ-23 in Leningrad manufactured 737 copies of the U-1. During the interim, Il'yushin proved to be so knowledgeable about aircraft that in 1922, at the end of the Russian Civil War, he was assigned to attend the Zhukovskii Military Air Academy. After graduation, he served as a member of the Red Air Force Scientific and Technical Committee and later would become its chief. In the 1930s Il'yushin left the military for a civilian post as deputy to Tupolev in the Department of Experimental Airplane Construction. He then moved on to head the USSR's Aviation Industry Administration. Like Tupolev, he would eventually earn international recognition for his multi-engine military and commercial aircraft. Shortly before World War II, Il'yushin was credited with the creation of the Soviet Union's most important single-engine ground attack plane, the BSh-2 (*Bronirovanni Shturmovik*

No. 2) armored assault aircraft. It later became the Il-2, with several numbered variations.[14]

Although all of these men showed great promise for the future development of aviation, they were not able to meet the immediate needs of the Red Air Force before 1924. Istrebitel' No. 1 and No. 2, from Polikarpov and Grigorovich, only exposed the limitations of the early indigenous fighter aircraft in Russia. The Grigorovich plane went into production in 1925 and the Polikarpov aircraft had a small production run; indeed, the Istrebitel' No. 1 remained so difficult to manage in flight that it was never actually assigned to a standard aviation detachment. Thus it should come as no surprise that the Soviet government used gold and hard currencies amassed from foreign trade to purchase a large number of aircraft abroad. In the early 1920s, for example, the Soviets signed a contract to buy fighter planes from the Netherlands. Anthony H. G. Fokker, who had built pursuit planes in Schwerin, Germany, during the Great War, returned to his homeland and established an aircraft factory in 1919 at the former Dutch Naval Air Service Base in Veere, Zeeland, before moving on to Amsterdam in 1924.[15]

There was a good reason for the Soviet government to want the Dutchman's aircraft. In the last year of the Great War, Fokker in Germany produced the D. VII pursuit plane, which became the scourge of the Allied skies in 1918. The D. VII approached being a sesquiplane, like a number of Nieuport models, which had a "one-and-a-half-wing," in which the lower wing was shorter and had a narrower chord; the smaller lower wing enhanced pilot vision below the aircraft during combat operations against enemy pursuits and ground targets. But the Fokker version did not quite qualify as a sesquiplane; its fuselage was a typical box-girder, braced by welded steel-tubes and covered by metals forward and plywood and fabric in the rear. Nevertheless, powered by a 185-hp BMW III (*Bayerische Motoren-Werke* No. 3) engine, the D. VII demonstrated excellent performance for its time, with a maximum speed of 117 mph, a combat range of 175 miles, and a service ceiling above 19,000 feet. Indeed, the airplane was so good that Article IV of the November 1918 armistice agreement specifically required that Germany turn over all of its D. VII–type aircraft to the victorious Allied Powers.[16]

After the armistice agreement, Fokker managed to smuggle 98 D. VII aircraft from Germany to his homeland (Netherlands). Approximately half of those planes went to the Dutch military, and Soviet Russia bought the remainder during the Russian Civil War. The Red Air Force was so pleased with the D. VII that in 1923 the Soviet government decided to sign a contract for 125 more fighter aircraft designed and constructed by the Fokker firm. This time the Soviets purchased the advanced and upgraded version of the D. VII—the new (and faster) D. XI. A true sesquiplane, the D. XI single-seat pursuit biplane was powered by a liquid-cooled Hispano-Suiza 8-cylinder engine. The fighter could reach a speed of 140 mph, travel a combat range of 273 miles, and reach an altitude above 20,000 feet. And it only required external wire braces for wheels. There were 2 interplane struts, but they were not needed since the wings were cantilevered. Only the V-struts were necessary to hold the top wing to the fuselage. Similar to earlier versions, the D. XI had 2 synchronized machine guns in front of the pilot cockpit.[17]

In 1924 the Soviet government purchased an additional 50 aircraft from Fokker, all of them a revised version of the D. XI. The improvement mainly focused on inserting a more powerful engine, a Napier Lion 12-cylinder, liquid-cooled motor generating a rating of 450 hp. The upgraded D. XIII could fly 160 mph and reach an altitude above 24,500 feet. What is fascinating about this Soviet purchase is that the new aircraft never entered the inventory of the Red Air Force. The aircraft were delivered to the Russian town of Lipetsk by the Voronezh River, a little more than 300 miles south-southeast of Moscow, but German pilots flew the planes and controlled the aerodrome there. In fact, Germany and Soviet Russia had quietly agreed to collaborate over their military aviation, but they delayed making it public until after the opening of the Genoa (Italy) Trade Conference on April 10, 1922. During conference sessions, it became clear that Germany and Russia were viewed and treated by the other countries as complete outcasts in the community of nations.[18] In all, some twenty-nine countries participated in the meeting—more if one counts members of the British Commonwealth.

One reason that Germany and Soviet Russia were viewed as pariahs—especially by Allied Powers—was that both countries still technically owed the victorious nations large reparations from the Great War. The Treaty of Versailles, which ended the Great War, had blamed Germany for starting the conflict, despite the prior actions of Serbia, the Austro-Hungarian Empire, and the Imperial Russian Empire. In 1921 the Reparations Commission handed Germany a bill for 33 billion dollars. For perspective, the entire U.S. government budget for 1922–1923 amounted to only 3.5 billion dollars. In short, Germany regarded the reparation demand as absolutely incredible. There were two main avenues for Germany to acquire the transferable hard currencies needed to pay the astonishingly colossal tab—foreign investment and favorable trade, where Germany made more money in selling goods abroad than it spent for goods purchased from other countries. The Genoa Trade Conference did nothing to help Germany secure these two sources of hard currency—a step that would have enabled the country to begin paying its enormous reparation debt.[19]

Although Russian debts from the Great War were large by the financial standards of 1922, they never came close to matching the reparations demanded of Germany. The United States had lent Great Britain and France 2.11 billion dollars during the war, but more than a billion dollars from these loans went to Imperial Russia, which used the money to buy wartime goods from Canada and America, including Curtiss aircraft and motors. Great Britain and France had assumed that Russia would eventually pay them back so that the two countries could settle the American loans. What is more, the two Allied Powers had supplied services and other war materiel, including a large number of aircraft and a substantial majority of the airplane engines that powered Russian-built aircraft. In addition, American banks directly loaned Russia millions of dollars, and in 1917 the United States opened a line of credit to the Provisional Government of 450 million dollars, of which 188 million dollars was spent before history swept that government away in November. As was the case involving Germany, the multi-nation conference in Genoa neither promoted trade nor foreign investment in Soviet Russia. The Soviets argued that they had supported peace, not conflict, and therefore had no responsibility for the Great War—and no obligation to pay Imperial Russian debts.[20]

Logically, the two outcast countries decided to work together to solve common problems. On April 16, 1922, representatives from Germany and

Soviet Russia met in the harbor town of Rapallo, just east of Genoa, and signed the Treaty of Rapallo, a Soviet-German agreement to resume diplomatic relations, to renounce their claims against each other for damages from the Great War, and to pledge an end to all discrimination in trade. The treaty promised to resolve issues related to aviation. As detailed earlier, the Soviet government faced serious problems in resurrecting the aircraft industry, which was merely one of a multitude of economic complications that authorities had to address. In seeking help from foreign industry to revive and develop the economy, the Soviet government issued a decree on November 23, 1920, that granted foreigners the opportunity of securing a Soviet contract for a concession. In essence, it was a lease for foreigners who had the skill or equipment to improve production in factories or mines. While the state continued to control the leased property, it guaranteed that for a stipulated time period the concessionaire would be free to administer the property without fear of confiscation. Soviet Russia would benefit by obtaining quality goods and pure minerals, securing new technologies and operational methods, and receiving a contractual percentage of the profits.[21]

One of the first concessionaires was an American businessman, Armand Hammer. As a direct result of conversations with Lenin, Hammer received a concession contract to operate an asbestos mine in Soviet Russia; he later secured a pencil-manufacturing concession. The rubles that he earned from these concessions had no value outside Russia, but he could use the internal currency to purchase quality works of art. He subsequently sold the paintings in the United States. The resulting cash helped place Hammer in a financial position that eventually enabled him to invest in and control California's Occidental Petroleum Corporation. On January 29, 1923, the German aircraft firm Junkers Flugzeuwerke AG signed a concession with the Soviet government that enabled the company to manufacture aircraft in Russia. Founded by Hugo Junkers, during the Great War the assembly plants had had different names, such as Junkers und Compagnie and Junkers-Fokker Werke AG. Even so, during that conflict, Junkers became the world's first company to design and build all-metal military planes, beginning with steel and moving to duralumin—an alloy of aluminum and other metals.[22]

In fact, by January 1923 the Junkers company was already operating an aircraft plant inside Soviet Russia under an agreement that it had signed with the Soviet government on February 6, 1922. From the Soviet perspective, the arrangement with Junkers actually supported Moscow's January 26, 1921, allotment of three million gold rubles for the advance of Soviet Russian aircraft industry. One of a series of instructions for the use of that gold was the creation of an indigenous all-metal aircraft production capacity. Only belatedly did the Junkers company realize that the whole purpose in granting Junkers to operate a subsidiary near Moscow was to jump-start the process of developing the independent production of Soviet aircraft. There was a Soviet rationale behind turning over to Junkers the empty Russo-Baltic Automobile Works factory in Fili, in the northeast Moscow region (*oblast'*): it would enable hundreds of Russian laborers to work alongside experienced German machinists and learn about the processes, methods, and technologies necessary to build quality all-metal military aircraft.[23]

The Germans had a different list of reasons for setting up shop in Soviet Russia—based on their country's response to the Treaty of Versailles. The treaty, signed on June 28, 1919, not only required that Germany turn over all military aircraft and related dirigibles and engines, it also prohibited the defeated country from maintaining any military flying equipment, aircraft, or training programs. Moreover, the German National Army could have no more than a hundred thousand men. The Inter-Allied Aviation Inspection Committee, a branch of the Inter-Allied Military Control Commission, had strong powers to inspect and enforce both the aviation sections of the Versailles Treaty and the decisions of the London Ambassadors' Conference of June 1920 and the Paris Agreements of January 1921. From the treaty and the decisions of these groups, German aviation was temporarily prohibited from designing and building any aircraft, including civilian models. The victorious Allied Powers did everything they could to prevent Germany from competing for sales with civilian aircraft manufacturers in Allied countries.[24]

The temporary enforcement of the prohibition against the German manufacture of civilian aeronautical materials was renewed several times until May 5, 1922, when it ended—at least in one sense. For perspective on what happened then, note again the February 1922 date on which Junkers had signed its postwar agreement with Soviet Russia to build aircraft there. If there were any doubt about the self-serving nature of Allied restrictions on German civilian aircraft, what happened on May 5, 1922, should have removed it. The Inter-Allied Aviation Inspection Committee changed its name to the Aviation Guarantee Committee and established strict standards for the production of German civilian aircraft. Under the new rules, such planes could not exceed a speed of 170 kmh (105 mph), a payload of 600 kg (1,323 pounds), a ceiling of 4,000 meters (13,123 feet), and a range of 300 km (186 miles). All were well below the performance figures then being achieved by most other aircraft built in 1922. As a result, the Aviation Guarantee Committee effectively prevented the Germans from designing and producing aircraft that could compete with airplanes manufactured in Allied countries such as France, Great Britain, or Italy.

But the real issue was that Germany was not permitted to produce a single military plane.[25] Because the Soviets wanted access to German aircraft and construction technologies, and the Germans wanted to build military aircraft and train military pilots in Soviet Russia, discussions on these issues began well before the Treaty of Rapallo. Some preliminary conversations got under way in April 1919, and by September 1921 Soviet Trade Commissar Krasin had launched secret negotiations in Berlin with the German military representative, Colonel Otto Hasse; Germany later established and funded a holding company that operated in Berlin and Moscow. The company oversaw the Junkers aircraft plant at Fili, shell factories at Petrograd (later Leningrad), Petrokrepost' and Tula, and a poison gas factory at Samara. A branch of the German army, independent of the German embassy in Moscow, administered German training sites in Soviet Russia, turning out armored vehicles near Kazan, gas-warfare equipment at Saratov, and aircraft at Lipetsk.[26]

To equip the Junkers plant in Fili, the company began to transfer some machinery from its Dessau headquarters by the Elbe River, southwest of Berlin, to the Russo-Baltic Automobile factory. After May 1922, Junkers could legitimately produce aeronautical parts and components for the Ju-20 floatplane, which it sent to Soviet Russia. The low-wing, two-place all-metal monoplane was a reconnaissance aircraft. During 1923 it was assembled, but not yet fully manufactured, at Fili. The Soviet government purchased twenty of these aircraft; they entered naval service as R-02 (*Razvedchik*-No. 02) airplanes, with units in the Baltic and Black Seas. The next plane, the Ju-21, was actually manufactured at Fili. Understandably, it bore a strong resemblance to the Ju-20, but looked different because it had standard landing gear instead of floats and it employed a high parasol wing. The monoplane also was a reconnaissance aircraft, but the second cockpit was armed with a machine gun. Some copies went to the Soviet military; others were sent to the German flight-training center at Lipetsk. Ju-21s expanded the German aerodrome's inventory of planes beyond the Fokker D. XIII aircraft.[27]

Over several years the 1,350 German and Russian workers built 170 aircraft. Powered by a 185 hp BMW IIIa engine, the Ju-20 and Ju-21 could attain a high speed of 146 mph—some 41 mph faster than the standard set for Germany by the Aviation Guarantee Committee from 1922 to 1925. The BMW IIIa was interesting because component parts went to Soviet Russia from Germany. At Fili the engine was assembled, not manufactured. Junkers originally had agreed to develop an engine plant at Fili, but it never happened. Just as before the Russian Revolution, most Soviet Russian motors continued in the 1920s to be of foreign design. The motor workshop, GAZ-2, merely occupied the French engine facility that had been formed in Moscow during the tsarist period. In the early 1920s the Aviation Industry Commissariat identified engines using "M" for motor (identical to the word in Russian). For example, M-1 represented the Gnome motor, while M-6 identified the Hispano-Suiza engine. Not until 1925 did Arkadi D. Shvetsov design a Soviet Russian radial powerplant at GAZ-4. Before World War II, most Soviet aircraft engines simply copied Western types.[28]

Although Soviet Russia was behind in developing indigenous aircraft motors, the government publicly supported military aviation—in sharp contrast to the domestic posture of other countries. In 1923 War Commissar Trotskii formed the Society of Friends of the Air Force (*Obshchestvo Druzei Vozdushnovo Flota*—ODVF), which soon opened five thousand branches across Soviet Russia and attracted more than a million members. The ODVF launched a nationwide campaign to collect rubles in order to build warplanes, and the Communist Party pressured workers to donate a day's wages to the society's efforts; the campaign raised enough money to finance the construction of more than one hundred combat aircraft. The society also publicized military aviation by arranging flight displays, factory visits, airplane lectures and exhibits, and plastering air force posters across the country. The exertions of ODVF were enhanced by *Osoaviakhim* (the Russian abridgement for the Association for the Promotion of Defense, Aviation, and Chemical Warfare), which served as a paramilitary organization for young people and advanced the use of parachutes, powered aircraft and gliders, wireless communication, and even the use of machine guns. In short, *Osoaviakhim* promoted military aviation among Russian youth and prepared them for service in the Red Air Force.[29]

CONCLUSION

★ ★ ★ ★ ★

FRANCE PLAYED A MAJOR role in the formation of Russia's development of military flight. It was France, not Russia, that invented the first forms of aerial flight in lighter-than-air craft and provided them to the world. Along with the armies of other powerful sovereign states, the Russian military, belatedly, created aerostatic balloon detachments that enhanced observation of static enemy ground positions, deploying them in conflicts against Japan (1904–1905), the Central Powers (1914–1917), and internal combatants (1918–1921). France also inspired Russia to consider a different, more controllable, and movable aerial technology. The 1909 flight of French pilot and aircraft designer Louis Blériot across the English Channel caught the attention of the Russian tsar's cousin and brother-in-law, Grand Duke Aleksandr Mikhailovich Romanov. His access to funds and to the hierarchy of the tsar and the military enabled him to purchase French aircraft and to send Russian officers to France for pilot training. The first airplane built in Russia in 1910 was a copy of the Blériot Model XI. Four of the first five major aircraft manufacturers established by Russians began by securing licenses to build French aircraft. And although Russia has been credited with developing as many as thirty-eight different indigenous aircraft during the Great War, the most successful

Russian military planes copied French and, to a lesser extent, British models. By 1917 more than 25 percent of all Russian combat planes were French Nieuports—either imported or Russian-built.

Even recent Russian sources admit that only two men deserve recognition as indigenous airplane designers before 1917—Igor Sikorsky, for his creation of the world's first four-engine reconnaissance-bomber, and Dimitrii Grigorovich, for his various flying boats. The most serious limitation to Russian military aviation had nothing to do with designing or constructing airframes. Well into the 1920s, the problem was Russia's inability to produce a decent engine capable of propelling an airframe into flight. Although Russia was filled with an abundance of natural resources, it suffered significantly from a weak industrial structure, an inadequate transportation system, limited technological progress, and a cumbersome and corrupt tsarist government. As a result, Russia's aviation development in the Great War could not match that of France, Germany, or Great Britain. Once again, France solved the aviation problem, with several French engine companies setting up subsidiaries in Russia to assemble the imported parts of aircraft motors. Between 1914 and 1917, those engines that were not assembled in Russia were imported mainly from France and from several Allied countries.

During the Great War there were positive and negative aspects to the use of military aircraft. The Imperial Russian Navy immediately accepted and quickly benefited from aerial reconnaissance. As early as 1913, the navy began to experiment with a forerunner of the modern aircraft carrier, which it used during the war in both the Baltic and Black Seas. By contrast, the Imperial Russian Army lagged behind. Several generals initially eschewed the use of reconnaissance aircraft, realizing the crucial value of airplanes only after months of war. Russia had its share of excellent pilots, but their aircraft always represented last year's model, whether imported or license-built in Russia. The country had an outstanding aerodynamicist and good ground schools, along with two quality flight-training centers. Unfortunately, flight training at other locations failed to match the relative excellence of Gatchina and Sevastopol'. Stavka assumed control over aviation and Grand Duke Aleksandr commanded the Directorate of the Military Aerial Fleet, but combat aviation activity usually was directed by corps generals and admirals commanding sea forces. The 5 major aircraft manufacturers increased monthly production from 35 planes in 1914 to 200 by 1916, yet France constructed 7,549 planes in 1 year compared with Russia's total of 5,550 for 3 years. Russia's important 4-engine reconnaissance-bomber was the Il'ia Muromets aircraft, but the constraints faced by Russian manufacturers prevented them from making any more than 22 of them available at any point in the war. Indigenous large bombers and flying boats, along with imported and license-built combat planes, played important roles over land and sea; nevertheless, either as a scouting network or destructive force they could not match the naval and army aircraft of World War II.

Finally, there were a series of issues that reduced or impaired the effectiveness of aviation during the Great War. The initial wartime draft of men for military service took skilled labor away from mines and factories, including aircraft workshops, resulting in poor craftsmanship that caused damage to planes and injury or death to pilots. There were sixteen different aircraft models and twelve different engines in the Russian inventory of planes at the start of the war. It was a horror story to maintain or repair all these different airplanes in the field. No one in aviation leadership thought either to focus on one or two types of high-performing aircraft or to insist on standardized parts and produce an abundance of replaceable parts for airplane repairs. The technology that enabled the Germans to manufacture a machine gun in 1915 that could fire synchronically through a rotating propeller on a smaller plane took well over a year for the Russians to produce. The same lag impeded radio communication between reconnaissance aircraft and ground forces, especially artillery. The application came late in the war and only among a very small number of selected units.

The 1917 Russian Revolution and Order No. 1 caused a sharp decline in military discipline and virtually ended the army's active engagement with the enemy. Aviation was an exception, however. All Russian military pilots had taken a solemn oath to protect the empire, even while it still was under Tsar Nikolai II, so most aviators continued to carry out their missions, whether they entailed scouting, bombing, or fighting the land, air, and sea forces of the Central Powers. Indeed, even after the Russian Revolution quite a few pilots achieved ace status as a result of aerial battles with German and Austrian aircraft. Russian naval pilots assisted in maintaining absolute control over the Black Sea. Russia's advantage began to erode in June, when a branch of the Soviet movement removed the Black Sea admiral and disarmed naval officers. In early July the Provisional Government, pressured by the Allies, attempted to conduct an offensive against German and Austrian troops. The army's air arm did everything necessary to prepare for battle. Pilots performed extensive reconnaissance duties to secure photographic information on the enemy's disposition of troops and to identify the best targets for Russian artillery; other fighter aircraft protected air scouts and attacked Austrian and German planes. Despite excellent aerial support and friendly and accurate artillery fire, the offensive collapsed when most ground soldiers refused to advance toward the enemy. At the same time, the Russian air war suffered from the economy breakdown, labor strikes, and troop desertions of men who previously had provided support and protection in the field of war. These factors posed terrible problems for aviation squadrons, many of which had ceased functioning by summer's end. Russia's last air battles were fought by navy pilots in October 1917, as part of an attempt to save the Gulf of Riga from enemy control; the effort failed when a significant German force on land, on sea, and in the air entered the gulf and forced the Russians to retreat.

With the Bolshevik phase of the revolution came the armistice and, later, the Treaty of Brest-Litovsk, which officially ended Soviet Russia's participation in the Great War. Meanwhile, the Bolsheviks, soon renamed Communists, waged class warfare against the bourgeoisie and the nobility. The hostilities led aircraft designers such as Igor Sikorsky and pilots like Alexander Seversky to flee from the former Russian Empire. A number of pilots from the middle and lower classes aligned themselves with the Soviet movement, while many pilots who were members of the nobility joined the anti-Red Whites, who sought an end to Communist rule. Several Allied countries supported the Whites, initially with hopes of reestablishing some type of military front against German and Austrian forces in the east in order to reduce the number of enemy troops and aircraft on the Western and Italian Fronts in 1918. White groups with pilots and aircraft began to emerge in North, Southeast European, and Siberian Russia. Inadvertently, the Czech-Slovak Legion's takeover of the Trans-Siberian Railroad enhanced the White movement and encouraged Allied military intervention.

CONCLUSION

With the emergence of an active civil war, the new Soviet government's military found its aviation resources spread thin among peripheral White areas of the former empire. The situation, along with British aid in North Russia in 1918 and Southeastern European Russia in 1919, actually allowed the White movement to obtain better aircraft, flown by more experienced ace pilots from the Great War. By contrast, the Soviets in the Red Air Force flew older, used, and obsolescent aircraft powered by ersatz fuel. The Soviets also inherited numerous types of aircraft, which made repairs seriously difficult. Along with the decline of Russian industry, including aircraft factories, replacement parts and new military planes were initially unavailable. The Soviets also had to reestablish pilot-training facilities to secure capable pilots; they used other measures to revive aircraft construction especially in Moscow. Through drastic steps and conscription, the Soviet government created a massive Red Army. Over time, the Army became more proficient as it forced Allied troops out of North Russia, defeated White forces—including Siberian units as well as those that formed east of Estonia—and severely pushed back the large number of White troops from South Russia.

In March 1920 a significant remnant of surviving White soldiers went by ship from the region north of the Caucasus to the Crimean Peninsula. Several aircraft squadrons accompanied the ground troops to this last major center of anti-Soviet forces. The very modest air force diminished in numbers in the weeks and months ahead. For various reasons Allied countries stopped supplying military hardware, including aircraft. Repairs to the heavily used White planes often required cannibalizing parts, propellers, and engines from other aircraft in the limited inventory. The Polish war with Soviet Russia goes a long way toward explaining how White partisans managed to exist for more than 6 months. The armistice between those 2 warring parties enabled the Red Army and Red Air Force to send a large number of troops and aircraft to retake the Crimea before the end of November. A brief extension of the civil war occurred in late winter of 1921, when sailors at the Kronshtadt Fortress on Kotlin Island revolted against the Soviet government. Two Red Army divisions and more than 130 Red Air Force combat missions crushed the rebellion.

Meantime, between 1918 and 1924 several key factors arose that help explain the success of Soviet military aviation during World War II, which amounted to a second version of the Great War. First, the foundation for improved military aircraft appeared in December 1918, with the formation by Nikolai Zhukovskii of what became the Central Aero-Hydrodynamic Institute. Best known by its Russian initials, TsAGI, the Institute became the most important and productive aeronautical research establishment in Soviet Russia—a role similar to the one played in the United States by the National Advisory Committee for Aeronautics, which became the National Aeronautics and Space Administration after 1958. Second, TsAGI and the Zhukovskii Military Air Academy served as birthplaces for future, even internationally known, military aircraft designers. Andrei Tupolev worked early on with Zhukovskii to create a new multi-engine aircraft to replace Sikorsky's Il'ia Muromets plane. In 1924 the young designer drafted the ANT-4, the world's first all-metal, twin-engine bomber. Large aircraft became his lifelong specialty. His Tu-2 twin-engine tactical bomber, designed in 1938, contributed to the Soviet Union's victory over the Germans during the Second World War.

Nikolai Grigorovich, who experimented at TsAGI for several years, designed the Istrebitel' No. 2, known as I-2. A graceful biplane with a welded steel-tube frame covered in wood, it became Soviet Russia's first indigenous military fighter to enter mass production in the 1920s, with more than two hundred copies built. Grigorovich's colleague, Nikolai Polikarpov, benefited from TsAGI's equipment, which helped him design Istrebitel' No. 1. The aircraft had a modern appearance with its low monowing, but the plane proved difficult to fly, and only a few of the I-1s were actually produced. Nevertheless, Polikarpov became famous for his many designs of fighter and twin-engine aircraft; in 1941, when Germany invaded the USSR, Polikarpov's I-16 accounted for 65 percent of the Soviet Union's total fighter aircraft. The last major Soviet aircraft designer, Sergei Il'yushin, actually received pilot training during the Great War. He later graduated from the Zhukovskii Military Air Academy. Il'yushin eventually headed the USSR's Aviation Industry Administration and received credit for the creation of the Shturmovik armored assault aircraft, which many consider to have been the most effective ground-attack plane of World War II.

Third, Soviet Russia and Weimar Germany were perceived as pariahs, especially by the former Allied Powers of the Great War. As a result, the two outcasts found common cause and formed official commercial and diplomatic relationships in the Treaty of Rapallo. Even before that treaty, the countries had begun negotiations leading to the Germans' secretly setting up an aircraft factory near Moscow. With the employment and interaction of Russians in a former automobile plant, Soviet Russia acquired both contemporary planes and the technology and methodology to build a large series of all-metal aircraft. Germans were able to construct military airplanes and train military pilots in the USSR, despite the restrictions in the Treaty of Versailles on what they could manufacture at home. Finally, in contrast to the governments of other countries, the Soviets did everything they could to raise awareness and active support of the citizenry for military aviation. The government or its air force sponsored aviation flights, lectures, exhibits, and factory visits and published and distributed air force posters. As early as 1923, the War Commissariat established the Society of Friends of the Air Force, creating thousands of branches and enlisting hundreds of thousands civilian members across Soviet Russia. In the long term, citizens donated millions of rubles to build combat aircraft. Thus, Soviet Russia took all necessary steps to garner the technology, design, advancement, and civic reinforcement for the production of modern military aircraft.

NOTES

CHAPTER ONE. PREPARING THE WAY

1. W. Bruce Lincoln, *Passage Through Armageddon* (New York: Oxford University Press, 1994), 17–40; Barbara W. Tuchman, *The Guns of August* (New York: Bantam Books, 1989), 91–104.
2. P. D. Duz', *Istoriia vozdukhoplavaniia i aviatsii v Rossii. Period do 1914 g.* (Moscow: Mashinostroenie, 1995), 2nd ed., 331–332; Alan Durkota, et al., *The Imperial Russian Air Service* (Mountain View, CA: Flying Machines Press, 1996), 24; Bob Woodling and Taras Chayka, *The Curtiss Hydroaeroplane* (Atglen, PA: Schiffer Publishing, Ltd., 2011), 91–102.
3. W. Bruce Lincoln, *The Romanovs* (Garden City, NY: Anchor Books, 1981), 142–142; Bruce W. Menning *Bayonets Before Bullets* (Bloomington: Indiana University Press, 1992), 152–199; Norman Stone, *The Eastern Front 1914–1917* (London, UK: Penguin Books, Ltd., 1998), 17–36; Simon Sebag Montefiore, *The Romanovs 1613–1918* (New York: Alfred A Knopf, 2016), 523.
4. Tom D. Crouch quotation from *The Bishop's Boys* (New York: W. W. Norton & Company, 1989), 299; George C. Herring, *From Colony to Superpower* (New York: Oxford University Press, 2008), 360–361; Edmund Morris, *Theodore Rex* (New York: Random House, 2001), 402–414 and 473; Valerie Moolman, *The Road to Kitty Hawk* (Alexandria, VA: Time-Life Books, 1980), 55, 61–61, and 92–94; Bill Gunston, *The Osprey Encyclopedia of Russian Aircraft 1875–1995* (London, UK: Osprey Aerospace, 1995), 7.
5. Heinz J. Nowarra and G. R. Duval, *Russian Civil and Military Aircraft 1884–1969* (London, UK: Fountain Press, 1971), 13–14.
6. Crouch, *Bishop's Boys*, 297–299 and 457; Herbert A. Johnson, *Wingless Eagle* (Chapel Hill: University of North Carolina Press, 2001), 112–113.
7. Tom D. Crouch, *Lighter Than Air* (Baltimore: Johns Hopkins University Press, 2009), 16–33; David R. Jones, "The Birth of the Russian Air Weapon 1909–1914," *Aerospace Historian* 21 (Fall 1974), 169; Walter J. Boyne, *The Influence of Air Power upon History* (Gretna, LA: Pelican Publishing Company, Inc., 2003), Appendix, 378; Menning *Bayonets Before Bullets*, 233.
8. Crouch, *Lighter Than Air*, 56; Lee Kennett, *The First Air War 1914–1918* (New York: The Free Press, 1991), 1–2; Heinz J. Nowarra and G. R. Duval, *Russian Civil and Military Aircraft 1884–1969* (London, UK: Fountain Press, 1971), 14; Lincoln, *Romanovs*, 587–589; Catherine Evtuhov, et al., *A History of Russia* (Boston: Houghton Mifflin Company, 2004), 443–445; Menning, *Bayonets Before Bullets*, 60–78; Montefiore,

Romanovs 1613–1918, 427–429; Nicholas N. Golovine, *The Russian Army in the World War* (New Haven: Yale University Press, 1931), 150.

9. A. K. Sorokin, otv. redactor, *Rossiia v Pervoi mirovoi voine 1914–1918* (Moskva: Politicheskaia entsiklopediia, 2014), 24; Nowarra and Duval, *Russian Civil and Military Aircraft 1884–1969*, 14; Nicholas N. Golovine, *The Russian Army in the World War* (New Haven: Yale University Press, 1931), 150.

10. Lincoln, *Passage Through Armageddon*, 130–131 and 154–155; Stone, *Eastern Front 1914–1917*, 148–149 and 174–175; Golovine, *Russian Army in the World War*, 222.

11. P(etr) D. Duz', *Istoriia vozdukhoplavaniia i aviatsii v SSSR. Period pervoi mirovoi voiny 1914–1918 g.g.* (Moscow: Oborongiz, 1960), 107–118.

12. Boris Sergievsky, *Airplanes, Women, and Song* (Syracuse: Syracuse University Press, 1999), 82; James K. Libbey, *Alexander P. de Seversky and the Quest for Air Power* (Washington, DC: Potomac Books, 2013), 9 and 33. The Russian aircraft manufacturer's name (Дух in Russian) is often rendered in English as Dux; however, the correct transliteration of the Russian letter "х" is "kh."

13. Crouch, *Lighter Than Air*, 70–77 and 82–84; Wilbur Cross, *Zeppelins of World War I* (New York: Barnes & Noble, 1991), 1–7.

14. Duz', *Istoriia vozdukhoplavaniia i aviatsii v SSSR*, 98–107; K. N. Finne, edited by Carl J. Borrow and Von Hardesty, *Igor Sikorsky* (Washington, D.C.: Smithsonian Institution Press, 1967), 87; Robert A. Kilmarx, "The Russian Imperial Air Forces of World War I.," *Air Power Historian*, 10 (July 1963), 90–91; Sorokin, otv. redactor, *Rossiia v Pervoi mirovoi voine 1914—1918*, 25; David R. Jones, "The Beginnings of Russian Air Power, 1907–1922," in Robin Higham and Jacob W. Kipp, eds., *Soviet Aviation and Air Power* (Boulder, CO: Westview Press, 1977), 16. James S. Corum and Richard R. Muller, *The Luftwaffe's Way of War: German Air Force Doctrine 1911–1945* (Baltimore: Nautical & Aviation Publishing Company of America, 1998), 3.

15. Christopher Chant, *Pioneers of Aviation* (New York: Barnes & Noble, 2001), 17–20; Nowarra and Duval, *Russian Civil and Military Aircraft 1884–1969*, 14; Libbey, *Alexander P. de Seversky and the Quest for Air Power*, 3; Duz', *Istoriia vozdukhoplavaniia i aviatsii v Rossii*, 297.

16. Charles Lindbergh quoted by A. Scott Berg, *Lindbergh* (New York: G. P. Putnam's Sons, 1998), 142; Tom D. Crouch, *Wings: A History of Aviation from Kites to the Space Age* (Washington, DC: Smithsonian National Air and Space Museum in association with W. W. Norton & Company, 2003), 89 and 108–114; Chant, *Pioneers of Aviation*, 34.

17. Alexander, Grand Duke of Russia, *Once a Grand Duke* (New York, Farrah & Rinehart, 1932), 25 and 80–87. Note: Grand Duke Aleksandr was lucky to escape Soviet Russia with his life thanks to a British war vessel in the Black Sea. Most of his papers remained behind. As a result, his autobiography lacks documentation and suffers from inaccuracies. See also: Menning, *Bayonets Before Bullets*, 233; Russell Miller, *The Soviet Air Force at War* (Alexandria, VA: Time-Life Books, 1985), 18; Gregory Vitarbo, *Army of the Sky* (New York: Peter Lang Publishing, 2012), 25; Nowarra and Duval, *Russian Civil and Military Aircraft 1884–1969*, 30; Jones, "Beginnings of Russian Air Power, 1907–1922," 17.

18. Alexander, *Once a Grand Duke*, 80–87, 120, 208, 228, and 237–238; Jones, "Beginnings of Russian Air Power, 1907–1922," Miller, *Soviet Air Force at War*, 18; Vitarbo, *Army of the Sky*, 27; Sorokin, otv. redactor, *Rossiia v Pervoi mirovoi voine 1914–1918*, 21.

19. Libbey, *Alexander P. de Seversky and the Quest for Air Power*, 7 and 9; Finne, *Igor Sikorsky*, 87; Gunston, *Osprey Encyclopedia of Russian Aircraft 1875–1995*, 2; Chant,

Pioneers of Aviation, 18, 21, 29, 32, and 35; Viktor Kulikov, "Aeroplanes of Lebedev's Factory," *Air Power History* 48 (Winter 2001), 6–7; James K. Libbey, "The Making of a War Hero," *Aviation History* 20 (March 2010), 54–59.

20. Libbey, *Alexander P. de Seversky and the Quest for Air Power*, 4–5; Durkota, et al., *Imperial Russian Air Service*, 268 and 336–337; Gunston, *Osprey Encyclopedia of Russian Aircraft 1875–1995*, 82–83.
21. Igor I. Sikorsky, *The Story of the Winged-S* (New York: Dodd Mead & Company, 1944), 1–139; Warren R. Young, *The Helicopters* (Alexandria, VA: Time-Life Books, 1982), 33–36 and see 79–81 for a discussion of Sikorsky's successful helicopter in the United States in 1940; James K. Libbey, "Sikorsky, Igor I. (1889–1972)," in Walter J. Boyne, ed., *Air Warfare*, vol. 2 (Santa Barbara, CA: ABC-CLIO, 2002), 567.
22. Nowarra and Duval, *Russian Civil and Military Aircraft 1884–1969*, 37–38; Gunston, *Osprey Encyclopedia of Russian Aircraft 1875–1995*, 1, 6, and 10; Durkota, et al., *Imperial Russian Air Service*, 303, 340–341, and 366–370.
23. Durkota, et al., *Imperial Russian Air Service*, 303; Vadim Mikheyev, *Sikorsky S-16* (Stratford, CT: Flying Machines Press, 1997), 4–10; M. S. Neshkin and V. M. Shabanov, comp., *Aviatory*, "Orlov, Ivan Aleksandrovich," (Moscow Rosspen), 217–218.
24. Nowarra and Duval, *Russian Civil and Military Aircraft 1884–1969*, 33–34; V. B. Shavrov, *Istoriia konstruktsii samoletov v SSSR do 1938g.* (Moscow: Mashinostroenie, 1986), 299–300 and see table 14 on p. 672.
25. Shavrov, *Istoriia konstruktsii samoletov v SSSR do 1938g.*, 302; John H. Morrow Jr., *The Great War in the Air* (Washington, DC: Smithsonian Institution Press, 1993), 294; Johnson, *Wingless Eagle*, 111; Duz', *Istoriia vozdukhoplavaniia i aviatsii v SSSR*, 272–295.
26. Roger E. Bilstein, *Flight in America* (Baltimore: Johns Hopkins University Press, 1994), 37; Crouch, *Wings*, 190–191' Finne, *Igor Sikorsky*, 35–36.
27. John W. R. Taylor, ed., *The Lore of Flight* (New York: Barnes & Noble, 1996), 50, 158, 185—186, and 189; Bill Gunston, *The Development of Piston Aero Engines* (Newbury Park, CA: Haynes North America, 1999), 110–112; Durkota, et al., *Imperial Russian Air Service*, 333; Sorokin, otv. redactor, *Rossiia v Pervoi mirovoi voine 1914–1918*, 22.
28. Sorokin, otv. redactor, *Rossiia v Pervoi mirovoi voine 1914*–1915, 22; Duz', *Istoriia vozdukhoplavaniia i aviatsii v SSSR*, 216–233; Nowarra and Duval, *Russian Civil and Military Aircraft 1884–1969*, 34; Sikorsky, *Story of the Winged-S*, 138; Theodore von Karman, *Aerodynamics* (Ithaca: Cornell University Press, 1954), 35–36.
29. V. P. Glushko, ed., "Zhukovskii, Nikolai Egorovich (1847–1921)," *Kosmonavtika entsiklopediia* (Moscow: Sovietskaia entsiklopediia, 1985), 114; Nowarra and Duval, *Russian Civil and Military Aircraft 1884–1969*, 11.
30. Grand Duke Alexander Mikhailovich quoted from *Once a Grand Duke*, 238; Duz', *Istoriia vozdukhoplavaniia i aviatsii v Rossii*, 334; Lincoln, *Romanovs*, 261–262. See also D. A. Kiuchariants, *Gatchina* (Leningrad: Lenizdat, 1990).
31. Miller, *Soviet Air Force at War*, 18; Kilmarx, "Russian Imperial Air Forces of World War I," 91; Alexander Riaboff, *Gatchina Days* (Washington, DC: Smithsonian Institution Press, 1986), 33–38; Duz', *Istoriia vozdukhoplavaniia i aviatsii v SSSR*, 256–257.
32. Riaboff, *Gatchina Days*, 38; Chant, *Pioneers of Aviation*, 18; Taylor, ed., *Lore of Flight*, 50. The Wright Model A is the common designation that commentators, but not the Wright Brothers, have given to the Wright Flyer after 1905 but before the Wrights manufactured and sold the Model B. The Model A was the first Flyer with seats. See James K. Libbey, "Flight Training and American Aviation Pioneers," *American Aviation Historical Society Journal* 49 (Spring 2004), 30.

33. Morrow, *Great War in the Air*, 12–13; Chant, *Pioneers of Aviation*, 18; Gunston, *Development of Piston Aero Engines*, 110–112; Alexander Riaboff, *Gatchina Days* (Washington, DC: Smithsonian Institution Press, 1986), 41–52.
34. Christopher C. Lovett, "Russian and Soviet Naval Aviation," in Higham, Greenwood, and Hardesty, eds., *Russian Aviation and Air Power in the Twentieth Century*, 108–110; Morrow, *Great War in the Air*, 48; Crouch, *Wings*, 144; Leonard E. Opdycke, *French Aeroplanes Before the Great War* (Atglen, PA: Schiffer Publishing, 1999), 270–271; John Barnhill, "Ely, Eugene (1886–1911)" in Boyne, ed., *Air Warfare*, Vol. 1, 193–194; Nowarra and Duval, *Russian Civil and Military Aircraft 1884–1969*, 31.

CHAPTER TWO. ON THE EVE AND START OF THE GREAT WAR

1. Jones, "Beginnings of Russian Air Power, 1907–1922," 17; Vitarbo, *Army of the Sky*, 181–182; René Greger, *The Russian Fleet 1914–1917* (London: Jan Allan, Ltd., 1972), 45; George Nekrasov, *North of Gallipoli* (Boulder, CO: East European Monographs, 1992), 14–15.
2. Vitarbo, *Army of the Sky*, 182–183; Jones, "Beginnings of Russian Air Power, 1907–1922," 17; Duz', *Istoriia vozdukhoplavaniia i aviatsii v Rossii*, 297. See also Aleksandr Nikolaevich Lapchinskii, *Vozduzhnaia razvedka* (Moskva: Gosvoenizdat N. K. O., 1938).
3. Neshkin and Shabanov, comp., *Aviatory*, "Rudnev, Evgenii Vladimirovich," 252.
4. Neshkin and Shabanov, Evgenii Vladimirovich," 251; Jones, "Beginnings of Russian Airpower, 1907–1922," 17.
5. Neshkin and Shabanov, comp., *Aviatory*, "Rudnev, Evgenii Vladimirovich," 251–252; Finne, *Igor Sikorsky*, 65 and 176–177.
6. Libbey, *Alexander P. de Seversky and the Quest for Air Power*, 11–12; Neshkin and Shabanov, comp., *Aviatory*, "Khodorovich, Viktor Antonovich," 296–297; Sorokin, otv. redactor, *Rossiia v Pervoi mirovoi voine 1914–1918*, 32; Jones, "Beginnings of Russian Air Power, 1907–1922," 22; Kilmarx, "Russian Imperial Air Forces of World War I," 91.
7. Riaboff, *Gatchina Days*, 16, 18, and 48; Jones, "Beginnings of Russian Air Power, 1907–1922."
8. Kilmarx, "Russian Imperial Air Forces of World War I," 91; Vitarbo, *Army of the Sky*, 176 and 218.
9. Jones, "Birth of the Russian Air Weapon 1909–1914," 171; Vitarbo, *Army of the Sky*, 82, Neshkin and Shabanov, comp., *Aviatory*, "Dybovskii, Viktor Vladimirovich" and "Nesterov, Petr Nikolaevich," 101–102 and 206–207; Michael A. J. Taylor, *The Aerospace Chronology* (London: Tri-Service Press, 1989), 42; Morrow, *Great War in the Air*, 32; Duz', *Istoriia vozdukhoplavaniia i aviatsii v SSSR*, 96 (Table 4).
10. Alexander, *Once a Grand Duke*, 237; Stone, *Eastern Front 1914–1917*, 28–35.
11. Jones, "Birth of the Russian Air Weapon 1909–1914," 171; Vitarbo, *Army of the Sky*, 82 and 86.
12. Neshkin and Shabanov, comp., *Aviatory*, "Prokof'ev (Severskii), Aleksandr Nikolaevich," 238–249; December 1940 Press Release, Ligue Internationale des Aviateurs, Library of Congress, American Institute of Aeronautics and Astronautics, Box 111, p. 1; Libbey, *Alexander P. de Seversky and the Quest for Air Power*, 7–12.

13. Felix Gilbert with David Clay Large, *The End of the European Era, 1890 to the Present* (New York: W. W. Norton & Company, 1991), 110–113; Sorokin, otv. redactor, *Rossiia v Pervoi mirovoi voine 1914–1918*, 25.
14. Sorokin, otv. redactor, *Rossiia v Pervoi mirovoi voine 1914–1918*, 22 and 24; Duz', *Istoriia vozdukhoplavaniia i aviatsii v SSSR*, 10; Jones, "Birth of the Russian Air Weapon 1909–1914," 171; Montefiore, *Romanovs 1613–1918*, 555–556.
15. Lincoln, *Passage Through Armageddon*, 34, 158, and 182–183; Stone, *Eastern Front 1914—917*, 185 and 221; Montefiore, *Romanovs 1613–1918*, 566.
16. Greger, *Russian Fleet 1914–1917*, 51; Nekrasov, *North of Gallipoli*, 66 and 87–91; Duz', *Istoriia vozdukhoplavaniia i aviatsii v Rossii*, 135–146; Vitarbo, *Army of the Sky*, 90; Sorokin, otv. redactor, *Rossiia v Pervoi mirovoi voine 1914–1918*, 22.
17. Nowarra and Duval, *Russian Civil and Military Aircraft 1884–1969*, 15–16 and 18; Sorokin, otv. redactor, *Rossiia v Pervoi mirovoi voine 1914–1918*, 25.
18. Duz', *Istoriia vozdukhoplavaniia i aviatsii v Rossii*, 48–54; Vitarbo, *Army of the Sky*, 96; Woodling and Chayka, *Curtiss Hydroaeroplane*, 95–97.
19. Woodling and Chayka, *Curtiss Hydroaeroplanes*, 96–97; Neshkin and Shabanov, comp., *Aviatory*, "Ragozin, Nikolai Aleksandrovich" and "Utgof, Viktor Viktorovich," 244–245 and 291–292.
20. Woodling and Chayka, *Curtiss Hydroaeroplane*, 100; Morrow, *Great War in the Air*, 81; Lee Kennett, *The First Air War 1914–1918* (New York: The Free Press, 1991), 20.
21. Grand Duke Alexander, quoted from *Once a Grand Duke*, 267; Corum and Muller, *Luftwaffe's Way of War*, 23–24; Duz', *Istoriia vozdukhoplavaniia i aviatsii v Rossii*, 411–412; Vitarbo, *Army of the Sky*, 208; Lincoln, *Romanovs*, 602; Lincoln, *Passage Through Armageddon*, 61–62.
22. Alexander, *Once a Grand Duke*, 267.
23. Jones, "Beginnings of Russian Air Power, 1907–1922," 23; Lovett, "Russian and Soviet Naval Aviation," 111; Sikorsky, *Story of the Winged-S*, 129–130; Von Hardesty, "Early Flight in Russia," in Higham, Greenwood, and Hardesty, eds., *Russian Aviation and Air Power in the Twentieth Century* (London: Frank Cass Publishers, 1998), 31.
24. A. J. P. Taylor, *The First World War* (New York: G. P. Putnam's Sons, 1980), 20–21; Tuchman, *Guns of August*, 34–43; Stone, *Eastern Front 1914–1917*, 40–41.
25. Gilbert with Large, *End of the European Era, 1890 to the Present*, 133; Taylor, *First World War*, 28–32; Lincoln, *Passage Through Armageddon*, 50.
26. Libbey, *Alexander P. de Seversky and the Quest for Air Power*, 7–8; Greger, *Russian Fleet 1914–1917*, 13–15.
27. Libbey, *Alexander P. de Seversky and the Quest for Air Power*, 7–8; Greger, *Russian Fleet 1914–1917*, 13–15; H(arold K.) Graf, *The Russian Navy in War and Revolution* (Honolulu: University Press of the Pacific, 2002), 4–7.
28. Tuchman, *Guns of August*, 297–301; Stone, *Eastern Front 1914–1917*, 44–54 and 70–91; Montefiore, *Romanovs 1613–1918*, 574–576; Winston S. Churchill, *The Unknown War* (New York: Charles Scribner's Sons, 1931), 174–190.
29. August G. Blume, "The Eastern Front," *Over the Front* 5 (Winter 1990), 341; Lincoln, *Passage Through Armageddon*, 126.
30. Neshkin and Shabanov, comp., *Aviatory*, "Nesterov, Petr Nikolaevich," 20620; Morrow, *Great War in the Air*, 83; For a more complete discussion of Nesterov see Konstantin I. Trunov, *Petr Nesterov* (Moskva: Sovietskaiia Rossiia, 1975).
31. Neshkin and Shabanov, comp., *Aviatory*, "Nesterov, Petr Nikolaevich," 20620; Morrow, *Great War in the Air*, 83; Hardesty, "Early Flight in Russia," 32.

32. Durkota, et al., *Imperial Russian Air Service*, 203–204; Nowarra and Duval, *Russian Civil and Military Aircraft 1884–1969*, 33. For a more complete discussion see Konstantin I. Trunov, *Petr Nesterov* (Moskva: Sovietskaiia Rossiia, 1975).

CHAPTER THREE. FALL 1914 CAMPAIGN

1. Neshkin and Shabanov, comp., *Aviatory*, "Tkachev, Viacheslav Matveevich," 281–284; Duz', *Istoriia vozdukhoplavaniia i aviatsii v Rossii*, 139; Vitarbo, *Army of the Sky*, 171–172.
2. Neshkin and Shabanov, comp., *Aviatory*, "Tkachev, Viacheslav Matveevich," 282; Stone, *Eastern Front 1914–1917*, 85–91; Golovine, *Russian Army in the World War*, 209–210 and 217–218.
3. Ibid.
4. Hindenburg quotation from Kennett, *First Air War 1914–1918*, 31. See also Martin van Creveld *The Age of Air Power* (New York: Public Affairs, 2011), 26; Taylor, *First World War*, 26 and 38; Tuchman, *Guns of August*, 205.
5. Lincoln, *Passage Through Armageddon*, 69–78; Stone, *Eastern Front 1914–1917*, 62–69; Golovine, *Russian Army in the World War*, 212–214.
6. Lincoln, *Passage Through Armageddon*, 69–78; Stone, *Eastern Front 1914–1917*, 62–69; Golovine, *Russian Army in the World War*, 212–214; Boris Simakov, "Russkaia aviatsiia v pervoi mirovi voine," *Vestnik vozdushnovo flota* 5 (1952), 89; Duz', *Istoriia vozdukhoplavaniia i aviatsii v SSSR*, 13; Sorokin, otv. redactor, *Rossiia v Pervoi mirovoi voine 1914–1918*, 29.
7. Blume, "Eastern Front," 341; John W. R. Taylor, ed., *Combat Aircraft of the World from 1909 to the Present* (G. P. Putnam's Sons, 1969), 582–583; Morrow, *Great War in the Air*, 81.
8. Taylor, *Lore of Flight*, 70–73; Kennett, *First Air War 1914–1918*, 100–101; Sorokin, otv. redactor, *Rossiia v Pervoi mirovoi voine 1914–1918*, 23.
9. Taylor, *Lore of Flight*, 189; Kennett, *First Air War 1914–1918*, 103–106.
10. Golovine, *Russian Army in the World War*, 150; Morrow, *Great War in the Air*, 47 and 83; Duz', *Istoriia vozdukhoplavaniia i aviatsii v SSSR*, 11.
11. Durkota, et al., *Imperial Russian Air Service*, 5.
12. Duz', *Istoriia vozdukhoplavaniia i aviatsii v SSSR*, 11; Kilmarx, "Russian Imperial Air Forces of World War I, " 91; Jones, "Beginnings of Russian Air Power, 1907–1922," 20.
13. James K. Libbey, *Russian-American Economic Relations* (Gulf Breeze, FL: Academic International Press, 1999), 59; Norman E. Saul, *War and Revolution* (Lawrence: University Press of Kansas, 2001), 12; Nowarra and Duval, *Russian Civil and Military Aircraft 1884–1969*, 49.
14. Kilmarx, "Russian Imperial Air Forces of World War I," 93; Duz', *Istoriia vozdukhoplavaniia i aviatsii v SSSR*, 221; Libbey, *Russian-American Economic Relations*, 59, 61, and 64–65; Saul, *War and Revolution*, 12 and 22–24.
15. Nekrasov, *North of Gallipoli*, 17–20; Tuchman, *Guns of August*, 186.
16. Gilbert with Large, *End of the European Era*, 106; Nekrasov, *North of Gallipoli*, 20–21. The Kaiser's full name was Friedrich Wilhelm Viktor Albert.
17. Greger, *Russian Fleet 1914–1917*, 45; Nekrasov, *North of Gallipoli*, 20–26.
18. Greger, *Russian Fleet 1914–1917*, 45; Nekrasov, *North of Gallipoli*, 20–26; Woodling and Chayka, *Curtiss Hydroaeroplane*, 100–101; Kennett, *First Air War 1914–1918*, 177.

19. Woodling and Chayka, *Curtiss Hydroaeroplane*, 101; Taylor, *First World War*, 77 and 79; Greger, *Russian Fleet 1914–1917*, 46.
20. Greger, *Russian Fleet 1914–1917*, 46; Woodling and Chayka, *Curtiss Hydroaeroplane*, 101; Nekrasov, *North of Gallipoli*, 29–34.
21. Nowarra and Duval, *Russian Civil and Military Aircraft 1884–1969*, 35–36; Woodling and Chayka, *Curtiss Hydroaeroplane*, 100; Finne, *Igor Sikorsky*, 184.
22. Sikorsky, *Story of the Winged-S*, 58–59; Frank J. Delear, *Igor Sikorsky* (New York: Dodd, Mead & Company, 1976), 37–38.
23. Sikorsky, *Story of the Winged-S*, 69–73; Finne, *Igor Sikorsky*, 34–37; Dorothy Cochrane, Von Hardesty, and Russell Lee, *The Aviation Careers of Igor Sikorsky* (Seattle: University of Washington Press for the National Air and Space Museum, 1989), 27 and 32.
24. Delear, *Igor Sikorsky*, 46–49; Cochrane, Hardesty, and Lee, *Aviation Careers of Igor Sikorsky*, 32–34.
25. Sikorsky, *Story of the Winged-S*, 90–92; Delear, *Igor Sikorsky*, 49–55; Cochrane, Hardesty, and Lee, *Aviation Careers of Igor Sikorsky*, 34–36.
26. Sikorsky, *Story of the Winged-S*, 95–101.
27. Hardesty, "Early Flight in Russia," 29–30; Finne, *Igor Sikorsky*, 47–55; Sikorsky, *Story of the Winged-S*, 102–117. Years later, Sikorsky claimed the round trip actually covered 1,600 miles. Because the four men had to deal with serious weather and visibility problems, his figure may actually be correct.
28. Hardesty, "Early Flight in Russia," 30; Gilbert with Large, *End of the European Era*, 113–119; Stone, *Eastern Front 1914–1917*, 42–43; Tuchman, *Guns of August*, 91–134.
29. Delear, *Igor Sikorsky*, 73–74; Sikorsky, *Story of the Winged-S*, 118–119.
30. Sikorsky, *Story of the Winged-S*, 122–123; Delear, *Igor Sikorsky*, 76–77; Taylor, *Lore of Flight*, 70–73.
31. Finne, *Igor Sikorsky*, 61–62; Neshkin and Shabanov, comp., *Aviatory*, "Rudnev, Evgenii Vladimirovich," 252.
32. Finne *Igor Sikorsky*, 52, 64–65, 176; Neshkin and Shabanov, comp., *Aviatory*, "Rudnev, Evgenii Vladimirovich," 252.

CHAPTER FOUR. NEW ROLES FOR AIRCRAFT IN 1915

1. Finne, *Igor Sikorsky*, 67–68; Neshkin and Shabanov, comp., *Aviatory*, "Pankrat'ev, Aleksei Vasil'evich," 222–224; Delear, *Igor Sikorsky*, 78.
2. Sikorsky, *Story of the Winged-S*, 124; Delear, *Igor Sikorsky*, 78.
3. Finne, *Igor Sikorsky*, 71–73; Sikorsky, *Story of the Winged-S*, 126–127; Lincoln, *Passage Through Armageddon*, 126.
4. Sikorsky, *Story of the Winged-S*, 126–130; Finne, *Igor Sikorsky*, 208–211; Delear, *Igor Sikorsky*, 208–211; Neshkin and Shabanov, comp., *Aviatory*, "Gorshkov, Georgii Georgievich," 88–90.
5. Neshkin and Shabanov, comp., *Aviatory*, "Gorshkov, Georgii Georgievich," 88–89.
6. Neshkin and Shabanov, comp., *Aviatory*, "Gorshkov, Georgii Georgievich," 88–89; Sikorsky, *Story of the Winged-S*, 127; Durkota, et al., *Imperial Russian Air Service*, 461; Neshkin and Shabanov, comp., *Aviatory*, "Kozakov, Aleksandr Aleksandrovich," 139–141.
7. Sikorsky, *Story of the Winged-S*, 127–129; Finne, *Igor Sikorsky*, 81; Taylor, ed., *Combat Aircraft of the World*, 135. One bomb did come close to injuring or killing Sikorsky.

8. Boyne, *Influence of Air Power upon History*, 76–77; Morrow, *Great War in the Air*, 91–92; Tom D. Crouch, *Wings: A History of Aviation from Kites to the Space Age* (New York: Smithsonian National Air and Space Museum in association with W. W. Norton & Company, 2003), 154–155.
9. Ibid.
10. Jones, "Beginnings of Russian Air Power, 1907–1922," 24; Neshkin and Shabanov, comp., *Aviatory*, "Voevodskii, Nikolai Stepanovich," 62–63.
11. Neshkin and Shabanov, comp., *Aviatory*, "Voevodskii, Nikolai Stepanovich," 63; Duz', *Istoriia vozdukhoplavaniia i aviatsii v SSSR*, 42–56; Morrow, *Great War in the Air*, 189; Libbey, *Alexander P. Seversky and the Quest for Air Power*, 33; William Green and Gordon Swanborough, *The Complete Book of Fighters* (New York: Barnes & Noble, 1998), 431–433; Ray Sanger, *Nieuport Aircraft of World War One* (Wiltshire, UK: Crowood Press, 2002), 117–118.
12. Lincoln, *Passage Through Armageddon*, 374–378; Durkota, et al., *Imperial Russian Air Service*, 59–61 and 117.
13. Graf, *Russian Navy in War and Revolution*, 34–36; Durkota, et al., *Imperial Russian Air Service*, 22, 25, and 268–270.
14. Greger, *Russian Fleet 1914–1917*, 14–16; Graf, *Russian Navy in War and Revolution*, 16–19.
15. Greger, *Russian Fleet 1914–1917*, 48; Nekrasov, *North of Gallipoli*, 59.
16. Neshkin and Shabanov, comp., *Aviatory*, "Utgof, Viktor Viktorovich," 291–292; Delear, *Igor Sikorsky*, 73 and 99–102; Cochrane, et al., *Aviation Careers of Igor Sikorsky*, 69–72; Sikorsky, *Story of the Winged-S*, 155–156; Greger, *Russian Fleet 1914–1917*, 14.
17. See "Summary of Events, 1914" in Greger, *Russian Fleet 1914–1917*, 15 and 47; Golovine, *Russian Army in the World War*, 212–218.
18. Lincoln, *Passage Through Armageddon*, 121–123; Sergievsky, *Airplanes, Women, and Song*, 56–58. Before becoming a pilot, Sergievsky was a battalion commander of ground troops in 1915 in the Carpathians.
19. Golovine, *Russian Army in the World War*, 127–151; Lincoln, *Passage Through Armageddon*, 106–107 and 121–122.
20. Robert A. Kilmarx, *A History of Soviet Air Power* (New York: Frederick A. Praeger, 1962), 23; Winston S. Churchill, *The Unknown War: The Eastern Front* (New York: Charles Scribner's Sons, 1931), 308–313; Stone, *Eastern Front 1914–1917*, 127–130.
21. Lincoln, *Passage Through Armageddon*, 125; Libbey, *Russian-American Economic Relations*, 59, 61, and 64–65; Duz', *Istoriia vozdukhoplavaniia i aviatsii v SSSR*, 59 and 221.
22. Churchill, *Unknown War*, 308–313; Stone, *Eastern Front 1914–1917*, 127–136; Lincoln, *Passage Through Armageddon*, 124–125.
23. Golovine, *Russian Army in the World War*, 222–224; Stone, *Eastern Front 1914–1917*, 137–143; Lincoln, *Passage Through Armageddon*, 126–129.
24. Stone, *Eastern Front 1914–1917*, 98 and 102; Lincoln, *Passage Through Armageddon*, 126. As the Russian Empire expanded in the 18th century, it acquired Baltic lands that contained a significant number of Germans. Many of them became officers in the Russian military and administrators in the government. Hence, a number of Russian officers, such as Admiral Nikolai von Essen and General Sergei Scheidemann, had distinctly German names but were loyal citizens of the Russian Empire.
25. Scheidemann quotation from Blume, "Eastern Front," 342, see also 341.

26. Neshkin and Shabanov, comp., *Aviatory*, "Kruten', Evgraf Nikolaevich," 156–157; Hardesty, "Early Flight in Russia," 32.
27. Neshkin and Shabanov, comp., *Aviatory*, "Grezo, Petr Petrovich" and "Kozakov, Aleksandr Aleksandrovich," 90 and 139–141. Some secondary sources list Kozakov as "Kazakov." His official military files in the Russian State History Archives spell his name with an "o" not an "a."
28. Neshkin and Shabanov, comp., *Aviatory*, "Kozakov, Aleksandr Aleksandrovich," 139–141; Miller, *Soviet Air Force at War*, 22 and 24; Hardesty, "Early Flight in Russia," 32.
29. Stone, *Eastern Front 1914–1917*, 171–172.
30. Stone, *Eastern Front 1914–1917*, 171–172; Greger, *Russian Fleet 1914–1917*, 17; Norman E. Saul, *Sailors in Revolt: The Russian Baltic Fleet in 1917* (Lawrence: The Regents Press of Kansas, 1978), 41.

CHAPTER FIVE. FLIGHT DURING THE GREAT RETREAT

1. Jacob W. Kipp, "Development of Naval Aviation, 1908–1975," in Higham and Kipp, eds., *Soviet Aviation and Air Power*, 140; Durkota, et al., *Imperial Russian Air Service*, 26; James K. Libbey, "The Making of a War Hero," *Aviation History* 20 (March 2010), 54–55.
2. Ibid.; Taylor, ed., *Combat Aircraft of the World*, 154–155; Greger, *Russian Fleet 1914–1917*, 19.
3. Graf, *Russian Navy in War and Revolution*, 33–36; Saul, *Sailors in Revolt*, 47.
4. Neshkin and Shabanov, comp., *Aviatory*, "Prokof'ev (Severskii), Aleksandr Nikolaevich," 238–240; Libbey, *Alexander P. de Seversky and the Quest for Air Power*, 8–12; Taylor, ed., *Combat Aircraft of the World*, 131–133.
5. Neshkin and Shabanov, comp., *Aviatory*, "Litvinov, Vladimir Aleksandrovich," 171–172. Lt. Litvinov was not a relative of Soviet Russia's famous diplomat and foreign minister, Maxim Litvinov. The diplomat was born Meer Genokh Moisevich Vallakh. See Hugh D. Phillips, *Between the Revolution and the West: A Political Biography of Maxim M. Litvinov* (Boulder, CO: Westview Press, 1992), 1.
6. Libbey, *Alexander P. de Seversky and the Quest for Air Power*, 17; Alexander P. de Seversky, "I Owe My Career to Losing a Leg," *Ladies' Home Journal* 61 (May 1944), 21.
7. Seversky quoted from de Seversky, "I Owe My Career to Losing a Leg," 22. At this time, Russian warships assigned to the Gulf of Riga were at the tip of Ösel Island's Sworbe Peninsula as the German navy attacked Windau north of Libau. See Graf, *Russian Navy in War and Revolution*, 48–49.
8. Greger, *Russian Fleet 1914–1917*, 49.
9. Greger, *Russian Fleet 1914–1917*, 50–52; Kilmarx, "Russian Imperial Air Forces of World War I," 93; Libbey, *Alexander P. de Seversky and the Quest for Air Power*, 14.
10. Greger, *Russian Fleet 1914–1917*, 51; Taylor, *First World War*, 98 and 100; Hugh Seton-Watson, *Eastern Europe Between the Wars 1918–1941* (Boulder, CO: Westview Press, 1986), 313.
11. Lincoln, *Passage Through Armageddon*, 182–183; Stone, *Eastern Front 1914–1917*, 190–191; Kennett, *First Air War 1914–1918*, 28; Duz', *Istoriia vozdukhoplavaniia i aviatsii v SSSR*, 107–118.
12. Nekrasov, *North of Gallipoli*, 66; Greger, *Russian Fleet 1914–1917*, 51; Seton-Watson, *Eastern Europe Between the Wars 1918–1941*, 313.

13. Finne, *Igor Sikorsky*, 87; Hardesty, "Early Flight in Russia," 34; Neshkin and Shabanov, comp., *Aviatory*, "Bashko, Iosif Stanislavovich," 36–38. Sikorsky gave a slightly earlier date for the bombing of the German train. See Sikorsky, *Story of the Winged-S*, 131.
14. Finne, *Igor Sikorsky*, 92; Neshkin and Shabanov, comp., *Aviatory*, "Bashko, Iosif Stanislavovich," 36–38; Sikorsky, *Story of the Winged-S*, 131.
15. Neshkin and Shabanov, comp., *Aviatory*, "Bashko, Iosif Stanislavovich," 36–38; Durkota, et al., *Imperial Russian Air Service*, 187–188.
16. Finne, *Igor Sikorsky*, 94–98; Neshkin and Shabanov, comp., *Aviatory*, "Bashko, Iosif Stanislavovich," 36–38.
17. Neshkin and Shabanov, comp., *Aviatory*, "Bashko, Iosif Stanislavovich" and "Smirnov, Mikhail Vladimirovich," 36–38 and 267–268; Finne, *Igor Sikorsky*, 96–98.
18. Deputy quoted from Lincoln, *Passage Through Armageddon*, 148; Finne, *Igor Sikorsky*, 96 and 98; Neshkin and Shabanov, comp., *Aviatory*, "Ozerskii, Dmitrii Alekseevich," 213–214; George F. Kennan, *Russia Leaves the War* (New York: Atheneum, 1967), 164.
19. Lincoln, *Passage Through Armageddon*, 144, 148, and 175–176; Finne *Igor Sikorsky*, 99.
20. Neshkin and Shabanov, comp., *Aviatory*, "Ozerskii, Dmitrii Alekseevich," and "Spasov, Mikhail Porfir'evich," 213–214 and 269; Finne, *Igor Sikorsky*, 98–99.
21. Finne, *Igor Sikorsky*, 54; Delear, *Igor Sikorsky*, 82–83.
22. Sikorsky, *Story of the Winged-S*, 132.
23. Golovine, *Russian Army in the World War*, 236–237; Lincoln, *Passage Through Armageddon*, 160–161, 170–171, and 179–182; Stone, *Eastern Front 1914-1917*, 187–193.
24. Neshkin and Shabanov, comp., *Aviatory*, "Ozerskii, Dmitrii Alekseevich" and "Spasov, Mikhail Porfir'evich," 213–214 and 269; Finne, *Igor Sikorsky*, 99–101.
25. Neshkin and Shabanov, comp., *Aviatory*, "Nizhevskii, Robert L'vovich," 207–209; Finne, *Igor Sikorsky*, 108.
26. Neshkin and Shabanov, comp., *Aviatory*, "Nizhevskii, Robert L'vovich" and "Pankrat'ev, Aleksei Vasil'evich," 207–209 and 222–224; Kilmarx, "Russian Imperial Air Forces of World War I," 94; Finne, *Igor Sikorsky*, 87 and 212.
27. Kilmarx, "Russian Imperial Air Forces of World War I," 94; Duz', *Istoriia vozdukhoplavaniia i aviatsii v SSSR*, 220–221; Morrow, *Great War in the Air*, 188–189.
28. Neshkin and Shabanov, comp., *Aviatory*, "Vakulovskii, Konstantin Konstantinovich," "Orlov, Ivan Aleksandrovich," and "Tomson, Petr-Eduard Martinovich," 54–55, 217–218, 285–286.
29. Neshkin and Shabanov, comp., *Aviatory*, "Vakulovskii, Konstantin Konstantinovich," 54–55.
30. Neshkin and Shabanov, comp., *Aviatory*, "Orlov, Ivan Aleksandrovich," 217–218; Duz', *Istoriia vozdukhoplavaniia i aviatsii v Rossii*, 331–333.
31. Neshkin and Shabanov, comp., *Aviatory*, "Tomson, Petr-Eduard Martinovich," 286–286.

CHAPTER SIX. THE HEIGHT OF THE AIR WAR

1. Morrow, *Great War in the Air*, 132–159 and 168; Crouch, *Wings*, 172–173; Stephen Budiansky, *Air Power: The Men, Machines, and Ideas that Revolutionized War, from Kitty Hawk to Iraq* (New York: Penguin Books, 2004), 68–69; James L. Stokesbury, *A Short History of Air Power* (New York: William Morrow and Company, 1986), 57–75; Kennett, *First Air War 1914-1918*, 179; Kilmarx, "Russian Imperial Air Forces of World War I," 94.

2. Lincoln, *Passage Through Armageddon*, 179–182 and 201–203; Golovine, *Russian Army in the World War*, 237–238; Dominic Lieven, *The End of Tsarist Russia* (New York; Viking, 2015), 344–345; Libbey, *Russian-American Economic Relations*, 62–64; Barbara Jackson Gaddis, "American Economic Interests in Russia, August 1914–March 1917," (M.A. thesis, University of Texas, 1966), 112.
3. Shavrov, *Istoriia konstruktsii samoletov v SSSR*, 300 and 302; Duz', *Istoriia vozdukhoplavaniia i aviatsii v SSSR*, 220–221, 232, and 296; Sorokin, otv. redactor, *Rossiia v Pervoi mirovoi voine 1914–1918*, 22; Morrow, *Great War in the Air*, 188–189.
4. Libbey, *Alexander P. de Seversky and the Quest for Air Power*, 179; Jones, "Beginnings of Russian Air Power, 1907–1922," 23; Sorokin, otv. redactor, *Rossiia v Pervoi mirovoi voine 1914–1918*, 26; Alexander, *Once a Grand Duke*, 272–274.
5. Grand Duke Aleksandr quoted twice from his autobiography, *Once a Grand Duke*, 274 and 277; Montefiore, *Romanovs 1613–1918*, 607–608; Lieven, *End of Tsarist Russia*, 88 and 347. Rasputin was poisoned, kicked, shot, and then thrown into a river.
6. Sorokin, otv. redactor, *Rossiia v Pervoi mirovoi voine 1914–1918*, 26; Delear, *Igor Sikorsky*, 83; Finne, *Igor Sikorsky*, 138–141; Nowarra and Duval, *Russian Civil and Military Aircraft 1884–1969*, 41 and 52; Jones, "Beginnings of Russian Air Power, 1907–1922," 25.
7. Finne, *Igor Sikorsky*, 123; Hardesty, "Early Flight in Russia," 34; Lincoln, *Passage Through Armageddon*, 155.
8. Neshkin and Shabanov, comp., *Aviatory*, "Konstenchik, Avenir Markovich," 149–150. For those unfamiliar with religions, Orthodox and Eastern Rite (Uniat) Roman Catholic priests marry and raise families. Western Roman Catholic priests may not marry.
9. Neshkin and Shabanov, comp., *Aviatory*, "Konstenchik, Avenir Markovich," 149–150 and "Iankovius, Viktor Fedorovich," 316–317; Sikorsky, *Story of the Winged-S*, 133–134.
10. Ibid.; Neshkin and Shabanov, comp., *Aviatory*, "Konstenchik, Avenir Markovich," 149–150; Finne, *Igor Sikorsky*, 128–129.
11. Neshkin and Shabanov, comp., *Aviatory*, "Konstenchik, Avenir Markovich," 149–150 and "Iankovius, Viktor Fedorovich," 316–317; Finne, *Igor Sikorsky*, 130.
12. Libbey, *Alexander P. de Seversky and the Quest for Air Power*, 19–23; Durkota, et al., *Imperial Russian Air Service*, 105; Neshkin and Shabanov, comp., *Aviatory*, "Prokof'ev (Severskii), Aleksandr Nikolaevich," 239.
13. Neshkin and Shabanov, comp., *Aviatory*, "Diterikhs, Vladimir Vladimirovich," 98–99; "Prokof'ev (Severskii), Aleksandr Nikolaevich," 239; Libbey, *Alexander P. de Seversky and the Quest for Air Power*, 23–24.
14. Ibid.
15. Bruce Robertson, *Air Aces of the 1914–1918 War* (Fallbrook, CA: Aero Publishers, 1964), 155; Gunston, *Osprey Encyclopedia of Russian Aircraft 1875–1995*, 82–85; Taylor, ed., *Combat Aircraft of the World*, 135; Libbey, *Alexander P. de Seversky and the Quest for Air Power*, 24.
16. Taylor, ed., *Combat Aircraft of the World*, 135; Durkota, et al., *Imperial Russian Air Service*, 276; Libbey, *Alexander P. de Seversky and the Quest for Air Power*, 24–25.
17. Durkota, et al., *Imperial Russian Air Service*, 105–106; Taylor, ed., *Combat Aircraft of the World*, 135; Libbey, *Alexander P. de Seversky and the Quest for Air Power*, 25.
18. Libbey, "The Making of a War Hero, 54–59; Finne, *Igor Sikorsky*, 126.
19. Finne, *Igor Sikorsky*, 126–128.
20. Finne, *Igor Sikorsky*, 132–133; Durkota, et al., *Imperial Russian Air Service*, 355–356; Igor' Andreev, *Boev'e samolety* (Moskva: Prostreks, 1994), 26, 32, and 36; Sorokin, otv. redactor, *Rossiia v Pervoi mirovoi voine 1914–1918*, 28.

21. Lincoln, *Passage Through Armageddon*, 239; Stone, *Eastern Front 1914-1917*, 238; Finne, *Igor Sikorsky*, 123-126; Durkota, et al., *Imperial Russian Air Service*, 205-214; Neshkin and Shabanov, comp., *Aviatory*, "Pankrat'ev, Aleksei Vasil'evich," 222-224; Sorokin, otv. redactor, *Rossiia v Pervoi mirovoi voine 1914-1918*, 28.
22. Green and Swanborough, *Complete Book of Fighters*, 413-414; Gunston, *Osprey Encyclopedia of Russian Aircraft 1875-1995*, 83; Sorokin, otv. redactor, *Rossiia v Pervoi mirovoi voine 1914-1918*, 32. For a detailed exploration of each aircraft type produced in Russia, see chapter 3 in Shavrov, *Istoriia konstruktsii samoletov v SSSR do 1938g*, 180-279.
23. Durkota, et al., *Imperial Russian Air Service*, 134-139; Sorokin, otv. redactor, *Rossiia v Pervoi mirovoi voine 1914-1918*, 32.
24. Neshkin and Shabanov, comp., *Aviatory*, "Kozakov, Aleksandr Aleksandrovich," 139-140; Green and Swanborough, *Complete Book of Fighters*, 82, 430-431, and 540; Lincoln, *Passage Through Armageddon*, 248-253.
25. Stone, *Eastern Front 1914-1917*, 237-256; Golovine, *Russian Army in the World War*, 241-242; Taylor, ed., *Combat Aircraft of the World*, 46; Green and Swanborough, *Complete Book of Fighters*, 82-83.
26. Lincoln, *Passage Through Armageddon*, 257; Duz', *Istoriia vozdukhoplavaniia i aviatsii v SSSR*, 112-119; Sorokin, otv. redactor, *Rossiia v Pervoi mirovoi voine 1914-1918*, 27 and 30.
27. Greger, *Russian Fleet 1914-1917*, 52-57; Nekrasov, *North of Gallipoli*, 70, 87, and 106; R. D. Layman, *Naval Aviation in the First World War* (Annapolis: Naval Institute Press, 1996), 94; Jacob W. Kipp, "Development of Naval Aviation, 1908-1975," in Higham and Kipp, eds., *Soviet Aviation and Air Power*, 140-141; Higham, et al., eds., *Russian Aviation and Air Power in the Twentieth Century*, 111.
28. Nekrasov, *North of Gallipoli*, 20-26; C. E. Black and E. C. Helmreich, *Twentieth Century Europe: A History* (New York: Alfred A. Knopf, 1963), 60.
29. Nekrasov, *North of Gallipoli*, 70; Lovett, "Russian and Soviet Naval Aviation, 1908-96," in Higham, et al., eds., *Russian Aviation and Air Power in the Twentieth Century*, 111; Greger, *Russian Fleet 1914-1917*, 53.
30. Nekrasov, *North of Gallipoli*, 77-86; Greger, *Russian Fleet 1914-1917*, 55.
31. Graf, *Russian Navy in War and Revolution*, 90-91; Nekrasov, *North of Gallipoli*, 107 and 128; Black and Helmreich, *Twentieth Century Europe*, 61; Greger, *Russian Fleet 1914-1917*, 57-60.

CHAPTER SEVEN. THE 1917 REVOLUTION IMPACTS SQUADRONS

1. Nekrasov, *North of Gallipoli*, 123; Greger, *Russian Fleet 1914-1917*, 61.
2. Lincoln, *Passage Through Armageddon*, 259-260; Duz', *Istoriia vozdukhoplavaniia i aviatsii v SSSR*, table 4 on 96; Nowarra and Duval, *Russian Civil and Military Aircraft 1884-1969*, 50; Taylor, ed., *Combat Aircraft of the World*, 126-127.
3. Finne, *Igor Sikorsky*, 144; Lincoln, *Passage Through Armageddon*, 259.
4. Lincoln, *Passage Through Armageddon*, 259; Evtuhov, et al., *History of Russia*, 574; Rex Wade, *The Russian Revolution, 1917* (New York: Cambridge University Press, 2005), 24-26.
5. Wade, *Russian Revolution, 1917*, 29-44; Lincoln, *Passage Through Armageddon*, 475-470; Robert Vincent Daniels, *The Conscience of the Revolution: Communist Opposition in*

Soviet Russia (New York: Simon and Schuster, 1969), 38 and 68; Montefiore, *Romanovs 1613-1918*, 613-619.
6. Evtuhov, et al., *History of Russia*, 539; Wade, *Russian Revolution, 1917*, 46 and 102-103; Edward Acton, *Russia: The Present and the Past* (London: Longman, 1986), 160-161; Michael Kort, *The Soviet Colossus: A History of the USSR* (Boston: Unwin Hyman, 1990), 88-91.
7. Lincoln, *Passage Through Armageddon*, 349-350 and 418-419; Evtuhov, et al., *History of Russia*, 594; Wade, *Russian Revolution, 1917*, 102-103; Saul, *War and Revolution*, 86.
8. Riaboff, *Gatchina Days*, 61-65; Lincoln, *Passage Through Armageddon*, 398; Stone, *Eastern Front 1914-1917*, 287.
9. Riaboff quoted from *Gatchina Days*, 73; see also 69-78; Neshkin and Shabanov, comp., *Aviatory*, "Chekhutov, Aleksandr Anikievich," 301-302.
10. Graf, *Russian Navy in War and Revolution*, 154-177; Saul, *Sailors in Revolt*, 81-114; Wade, *Russian Revolution, 1917*, 110.
11. Sanger, *Nieuport Aircraft of World War One*, 117; Gunston, *Osprey Encyclopedia of Russian Aircraft*, 84; Libbey, *Alexander P. de Seversky and the Quest for Air Power*, 32.
12. Sanger, *Nieuport Aircraft of World War One*, 117; Norman Franks, *Nieuport Aces of World War I* (Oxford, UK: Osprey Publishing, 2000), 8-9; Taylor, ed., *Combat Aircraft of the World*, 114-115; Libbey, *Alexander P. de Seversky and the Quest for Air Power*, 32-33.
13. Saul, *Sailors in Revolt*, 76; Riaboff, *Gatchina Days*, 70; Greger, *Russian Fleet 1914-1917*, 61-63; Nekrasov, *North of Gallipoli*, 123-128. Graf, who served as a naval officer on a destroyer during World War I and the revolution, claims that Nepenin's murderer wore the uniform of a navy petty officer, but may have been a hired killer by Nepenin's replacement. See Graf, *Russian Navy in War and Revolution*, 120-121.
14. Morrow, *Great War in the Air*, 258; Finne, *Igor Sikorsky*, 142-151; Sikorsky, *Story of the Winged-S*, 139.
15. Wade, *Russian Revolution, 1917*, 90-101 and 135-141; Acton, *Russia*, 179; Evtuhov, et al., *History of Russia*, 598-599.
16. Evtuhov, et al., *History of Russia*, 598; Wade, *Russian Revolution, 1917*, 172-182.
17. Durkota, et al., *Imperial Russian Air Service*, "Donat Aduiovich Makienok," 86-89, and "Vasilii Ivanovich Ianchenko," 135-139; Neshkin and Shabanov, comp., *Aviatory*, "Argeev, Pavel Vladimirovich," 25-26, "Gil'sher, Georgii (Urii) Vladimirovich," 75-76, "Kozakov, Aleksandr Aleksandrovich," 139-141, "Kruten', Evgraf Nikolaevich," 156-157, and "Orlov, Ivan Aleksandrovich," 217-218.
18. Lincoln, *Passage Through Armageddon*, 409; Sorokin, otv. redactor, *Rossiia v Pervoi mirovoi voine 1914-1918*, 31.
19. Durkota, et al., *Imperial Russian Air Service*, "Ivan Aleksandrovich Loiko," 84-85, and "Aleksandr Mikhailovich Pishvanov," 95-97; Neshkin and Shabanov, comp., *Aviatory*, "Smirnov, Ivan Vasil'evich," 265-266, "Strzhizhevskii, Vladimir Ivanovich," 273-275, and Suk, Grigorii Eduardovich," 277-278; Sorokin, otv. redactor, *Rossiia v Pervoi* mirovoi voine 1914-1918, 31.
20. Lincoln, *Passage Through Armageddon*, 411; Sorokin, otv. redactor, *Rossiia v Pervoi mirovoi voine 1914-1918*, 31; Neshkin and Shabanov, comp., *Aviatory*, "Vakulovskii, Konstantin Konstantinovich," 54-55.
21. Wade, *Russian Revolution, 1917*, 182-183; Lincoln, *Passage Through Armageddon*, 409-411.
22. Finne, *Igor Sikorsky*, 151; Sorokin, otv. redactor, *Rossiia v Pervoi mirovoi Voine 1914-1918*, 28; Lincoln, *Passage Through Armageddon*, 485-500. The Russian Air Force

Museum is located 38 kilometers [23.5 miles] east of Moscow at Monino, which has a copy of the Il'ia Muromets.
23. Finne, *Igor Sikorsky*, 151; Sorokin, otv. redactor, *Rossiia v Pervoi mirovoi Voine 1914–1918*, 28; Lincoln, *Passage Through Armageddon*, 485–500; Neshkin and Shabanov, comp., *Aviatory*, "Pankrat'ev, Aleksei Vasil'evich," 222–224.
24. Saul, *War and Revolution*, 140–141; Evtuhov, et al., *History of Russia*, 598; Sanger, *Nieuport Aircraft of World War One*, 117; Kennan, *Russia Leaves the War*, 37; Libbey, *Alexander P. de Seversky and the Quest for Air Power*, 34.
25. Seversky quoted from Libbey, *Alexander P. de Seversky and the Quest for Air Power*, 35, but see also 34 and 36.
26. Seversky quoted from Libbey, *Alexander P. de Seversky and the Quest for Air Power*, 35; Sanger, *Nieuport Aircraft of World War One*, 118; Durkota, et al., *Imperial Russian Air Service*, 280; Morrow, *Great War in the Air*, 96–97; Stone, *Eastern Front 1914–1917*, 282; Marion E. Grambow, ed., "De Seversky, Maj. Alexander P.," in *Who's Who in World Aviation and Astronautics*, vol. 2 (Washington: American Aviation Publications, 1958), 116–117. After Seversky became commander of the Second Pursuit Squadron, he had an accident on his motorcycle that left him bruised and with fractures to his good leg. Thus he had lost his command position. His success in learning to fly the Nieuport 17, then training others, and delivering a productive pep talk in the factory earned him the opportunity to return to his previous position. See Libbey, *Alexander P. de Seversky and the Quest for Air Power*, 30–35.
27. Durkota, et al., *Imperial Russian Air Service*, "Mikhail Ivanovich Safonov," 98–101; Lincoln, *Passage Through Armageddon*, 418.
28. Sorokin, otv. redactor, *Rossiia v Pervoi mirovoi voine 1914–1918*, 31; Greger, *Russian Fleet 1914–1917*, 27; Saul, *Sailors in Revolt*, 158.
29. Greger, *Russian Fleet 1914–1917*, 27; Saul, *Sailors in Revolt*, 158–159; Graf, *Russian Navy in War and Revolution*, 147–148; Libbey, *Alexander P. de Seversky and the Quest for Air Power*, 37–38.
30. Greger, *Russian Fleet 1914–1917*, 27; Saul, *Sailors in Revolt*, 159; Graf, *Russian Navy in War and Revolution*, 152; Durkota, et al., *Imperial Russian Air Service*, 108; Libbey, *Alexander P. de Seversky and the Quest for Air Power*, 38–39.
31. James Farber, "Major de Seversky–Engineer," *Popular Aviation* 17 (August 1935), 87–88, 116, and 132; Libbey, *Alexander P. de Seversky and the Quest for Air Power*, 39–40; Durkota, et al., *Imperial Russian Air Service*, "Mikhail Ivanovich Safonov," 98–101.
32. Farber, "Major Seversky—Engineer," 116; Graf, *Russian Navy in War and Revolution*, 151–153; Greger, *Russian Fleet 1914–1917*, 29–30; Alexander P. de Seversky, *Air Power* (New York: Simon & Schuster, 1950), Appendix 359; "Maj. Alexander P. de Seversky," in Biographies of Pioneer Aviators, Orvil A. Anderson Collection, U.S. Air Force Historical Research Agency, 168.7006-47, p. 2; Durkota, et al., *Imperial Russian Air Service*, 476.

CHAPTER EIGHT. REDS VERSUS WHITES

1. Durkota, et al., *Imperial Russian Air Service*, "Mikhail Ivanovich Safonov," 97; Neshkin and Shabanov, comp., *Aviatory*, "Prokof'ev (Severskii), Aleksandr Nikolaevich," 238; Graf, *Russian Navy in War and Revolution*, 173; Evan Mawdsley, *The Russian Civil War* (New York: Pegasus Books, 2005), 34.

2. Mawdsley, *Russian Civil War*, 27-29; Evtuhov, et al., *History of Russia*, 504-505; Durkota, et al., *Imperial Russian Air Service*, "Mikhail Ivanovich Safonov," 100.
3. Durkota, et al., *Imperial Russian Air Service*, "Mikhail Ivanovich Safonov," 101; Seversky, *Air Power*, Appendix, 359; "Maj. Alexander P. de Seversky, " in Biographies of Pioneer Aviation, Orvil A. Anderson Collection, U.S. Air Force Historical Research Agency, 168.7006-47, 2.
4. John Lewis Gaddis, *Russia, the Soviet Union and the United States* (New York: McGraw-Hill, 1998), 71; Adam B. Ulam, *Expansion and Coexistence: The History of Soviet Foreign Policy 1917-67* (New York: Frederick A. Praeger, 1968), 68-69; George F. Kennan, *The Decision to Intervene* (New York: Atheneum, 1967), 43-46.
5. "Report of the November 1940 Meeting," *The Adventurer* (December 1940), 8; Swanee Taylor, "Seversky: An Ace's Place Is in the Air," *New York World Telegram Metropolitan Weekend Magazine* (December 12, 1936), 16; "Russian Flyer with One Leg Seeks U.S. Job," *San Francisco Examiner*, April 22, 1918; Durkota, et al., *Imperial Russian Air Service*, "Mikhail Ivanovich Safonov," 101. The Soviet government had approved the peace treaty with the Central Powers by the time Seversky reached Tokyo. Nevertheless, the Russian Embassy temporarily made him a deputy naval attaché, even though the Russian naval aviation mission ended with the treaty.
6. Libbey, *Alexander P. de Seversky and the Quest for Air Power*, 49-266; Durkota, et al., *Imperial Russian Air Service*, "Mikhail Ivanovich Safonov," 101; Ulam, *Expansion and Coexistence*, 33; Evtuhov, et al., *History of Russia*, 609; Robert H. Donaldson and Joseph L. Nogee, *The Foreign Policy of Russia: Changing Systems, Enduring Interests* (New York: M. E. Sharpe, 2002), 50-52.
7. Mawdsley, *Russian Civil War*, 3-4; Wade, *Russian Revolution, 1917*, 248; Evtuhov, et al., *History of Russia*, 602-603; James K. Libbey, *Alexander Gumberg and Soviet-American Relations 1917-1933* (Lexington: University Press of Kentucky, 1977), 21 and 23; Kennan, *Russia Leaves the War*, 72; Daniels, *Conscience of the Revolution*, 63.
8. Wade, *Russian Revolution, 1917*, 248-251; Kennan, *Russia Leaves the War*, 74-75; Libbey, *Alexander Gumberg and Soviet-American Relations 1917-1933*, 23; Acton, *Russia*, 186-187.
9. W. Bruce Lincoln, *Red Victory: A History of the Russian Civil War* (New York: Da Capo Press, 1999), 100-101, 128, and 156-157; Daniels, *Conscience of the Revolution*, 63-69 and 92; Mawdsley, *Russian Civil War*, 40; Ulam *Expansion and Coexistence*, 88; Kennan, *Decision to Intervene*, 434 and 462.
10. Mawdsley, *Russian Civil War*, 49; Lincoln, *Red Victory*, 100-101 and 120-121.
11. Evtuhov, et al., *History of Russia*, 611; Kennan, *Decision to Intervene*, 414-415; Lincoln, *Red Victory*, 77; Mawdsley, *Russian Civil War*, 20 and 167.
12. George Stewart, *The White Armies of Russia: A Chronicle of Counter-Revolution and Allied Intervention* (New York: Russell & Russell, 1970), 25-39; General A. I. Denikin, *The White Army* (Cambridge, UK: Ian Faulkner Publisher, 1992), 14-18; Mawdsley, *Russian Civil War*, 12 and 20; Lincoln, *Red Victory*, 80-82; Neshkin and Shabanov, comp., *Aviatory*, "Tkachev, Viacheslav Matveevich," 281-284; Durkota, et al., *Imperial Russian Air Service*, "Viacheslav Matveevich Tkachev," 234-238. For some background information on Vrangel' and his take on the Revolution, see his autobiography: General Petr N. Vrangel', *Always With Honour* (New York: Robert Speller & Sons, 1957), 3-50.
13. Denikin, *White Army*, 19, 73-81, 84, 152-173; Lincoln, *Red Victory*, 84-88; Mawdsley, *Russian Civil War*, 19-21; Vrangel', *Always With Honour*, 51-78; Stewart, *White Armies of Russia*, 40-41.

14. Alexander Boyd, *The Soviet Air Force Since 1918* (New York: Stein and Day, 1977), 2–3; Robert A. Kilmarx, *A History of Soviet Air Power* (New York: Frederick A. Praeger, 1962), 35–36; D. Fedotoff White, *The Growth of the Red Army* (Princeton: Princeton University Press), 32; Kenneth R. Whiting, *Soviet Air Power* (Boulder, CO: Westview Press, 1986), 4.
15. Robert Jackson, *The Red Falcons: The Soviet Air Force in Action, 1919–1969* (London: Clifton Books, 1970), 10–11; Gunston, *Osprey Encyclopedia of Russian Aircraft*, "Dimitrii Pavlovich Grigorovich," 82–95, see especially 86.
16. Mawdsley, *Russian Civil War*, 38 and 200; Kilmarx, *History of Soviet Air Power*, 38–39.
17. Kilmarx, *History of Soviet Air Power*, 37–38; Neshkin and Shabanov, comp., *Aviatory*, "Belousovich, Nikolai Ivanovich," 41–42.
18. Neshkin and Shabanov, comp., *Aviatory*, Bashko, "Iosif Stanislavovich," 36–38; Durkota, et al., *Imperial Russian Air Service*, "Iosif Stanislavovich Bashko," 187–196.
19. Neshkin and Shabanov, comp., *Aviatory*, "Bashko, Iosif Stanislavovich, 36–38; Mawdsley, *Russian Civil War*, 226; Whiting, *Soviet Air Power*, 56.
20. Kilmarx, *History of Soviet Air Power*, 39; Whiting, *Soviet Air Power*, 4–8; Gunston, *Osprey Encyclopedia of Russian Aircraft*, 39, 86, 132, and 378. Both army and navy were ultimately under the Revolutionary Military Council chaired by Lenin. See James Bunyan, *International Civil War and Communism in Russia, April–December 1918* (Baltimore: Johns Hopkins Press, 1936), 515.
21. Mawdsley, *Russian Civil War*, 24–26; Neshkin and Shabanov, comp., *Aviatory*, "Vologodtsev (Grinshteii), Lev Konstantinovich," 64–65. In 1916 Lev changed his family name to Grinshteii.
22. Neshkin and Shabanov, comp., *Aviatory*, "Vologodtsev (Grinshteii), Lev Konstantinovich," 64–65; Lincoln, *Red Victory*, 305–308.
23. Kennan, *Russia Leaves the War*, 371; Lincoln, *Red Victory*, 93; Stewart, *White Armies of Russia*, 96–99.
24. Stewart, *White Armies of Russia*, 99–106; Evtuhov, et al., *History of Russia*, 611–612; Kennan, *Decision to Intervene*, 136–165; Lincoln, *Red Victory*, 94–95.
25. Kennan, *Decision to Intervene*, 18 and 107–135; Libbey, *Alexander Gumberg and Soviet-American Relations 1917–1933*, 43; Stewart, *White Armies of Russia*, 83–86.
26. Kennan, *Decision to Intervene*, 417–429; Neshkin and Shabanov, comp., *Aviatory*, "Kozakov, Aleksandr Aleksandrovich," 139–141; "Modrakh, Sergei Karlovich," 191–192; "Sveshnikov, Aleksandr Nikolaevich," 260–261; Shebalin, Sergei Konstantinovich," 304–306.
27. Kennan, *Decision to Intervene*, 378–379; George Schatunowski, "The Civil War to the Second World War," in Asher Lee, ed., *The Soviet Air and Rocket Forces* (New York: Frederick A. Praeger, 1959), 23; H. F. Griffith, *RAF in Russia* (London: Hammond and Company, 1942), 6; Neshkin and Shabanov, comp. *Aviatory*, "Kozakov, Aleksandr Aleksandrovich," 139–141, see also 191–192, 260–261, and 304–306.
28. Neshkin and Shabanov, comp., *Aviatory*, "Kozakov, Aleksandr Aleksandrovich," 139–141; Taylor, ed., *Combat Aircraft of the World*, 339–342; James Streckfuss, "De Havilland Aircraft" and "Royal Aircraft Factory," in Boyne, ed., *Air Warfare*, Vols. 1 and 2, 339–342 and 534–535.
29. Mawdsley, *Russian Civil War*, 49–51; Stewart, *White Armies of Russia*, 91–95; Kilmarx, *History of Soviet Air Power*, 51; V. S. Shumikhin, *Sovietskaia voennaia aviatsiia: 1917–1941* (Moskva: Nauka, 1986), 60.

30. Stewart, *White Armies of Russia*, 186-187; Kilmarx, *History of Soviet Air Power*, 51; Shumikhin, *Sovetskaia voennaia aviatsiia*, 61; Streckfuss, "Caudron Aircraft (Early Years)," in Boyne, ed., *Air Warfare*, vol. 1, 121.
31. Kilmarx, *History of Soviet Air Power*, 51; Leonid I. Strakhovsky, *Intervention at Archangel* (Princeton: Princeton University Press, 1944), 161; Lincoln, *Red Victory*, 273-275; Kennan, *Decision to Intervene*, 425; Neshkin and Shabanov, comp., *Aviatory*, "Sveshnikov, Aleksandr Nikolaevich," 260-261.

CHAPTER NINE. AVIATION AND THE CIVIL WAR

1. Neshkin and Shabanov, comp., *Aviatory*, "Kozakov, Aleksandr Aleksandrovich," 139-141, and "Belousovich, Nikolai Ivanovich," 41-42; Kilmarx, *History of Soviet Air Power*, 51; W[illiam] E[dmund] Ironside, *Archangel, 1918-1919* (London: Constable & Company, 1953), 157.
2. Green and Swanborough, *Complete Book of Fighters*, 536-537; Taylor, ed., *Combat Aircraft of the World*, 423-424.
3. Boyd, *Soviet Air Force Since 1918*, 3; Christopher H. Sterling, "Fairey Aircraft," in Boyne, ed., *Air Warfare*, vol. 1, 213-214; Green and Swanborough, *Complete Book of Fighters*, 533; Taylor, ed., *Combat Aircraft of the World*, 416; Andreev, *Boev'e samolety*, 16.
4. Ibid.: Boyd, *Soviet Air Force Since 1918*, 240; Green and Swanborough, *Complete Book of Fighters*, 12-13.
5. Whiting, *Soviet Air Power*, 5; Lee, *Soviet Air Force*, 24; Shumikhin, *Sovetskaia voennaia avciatsiia: 1917-1941*, 59.
6. Lincoln, *Red Victory*, 240-243; Mawdsley, *Russian Civil War*, 108-109; Kennan, *Decision to Intervene*, 408.
7. Ibid; Denikin, *White Army*, 224 and 232; Stewart, *White Armies of Russia*, 239-245.
8. Lincoln, *Red Victory*, 282-283; Mawdsley, *Russian Civil War*, 157.
9. Stewart, *White Armies of Russia*, 192-202; Lincoln, *Red Victory*, 281; Neshkin and Shabanov, comp., *Aviatory*, "Kozakov, Aleksandr Aleksandrovich," 139-141; Durkota, et al., *Imperial Russian Air Service*, "Aleksandr Aleksandrovich Kozakov," 58-71; Denikin, *White Army*, 273-274.
10. Durkota, et al., *Imperial Russian Air Service*, "Aleksandr Aleksandrovich Kozakov," 58-71; Neshkin and Shabanov, comp., *Aviatory*, "Belousovich, Nikolai Ivanovich," 41-42 and "Modrakh, Sergei Karlovich," 191-192.
11. Denikin, *White Army*, 230 and 232; Kilmarx, *History of Soviet Air Power*, 54; Chant, *Pioneers of Aviation*, 52-53; Taylor, ed., *Combat Aircraft of the World*, 482-484.
12. Mawdsley, *Russian Civil War*, 106-111; Lincoln, *Red Victory*, 238-248; Kilmarx, *History of Soviet Air Power*, 53.
13. Denikin, *White Army*, 226; Kennan, *Decision to Intervene*, 350; Kilmarx, *History of Soviet Air Power*, 53-54; Lincoln, *Red Victory*, 249-250.
14. Stewart, *White Armies of Russia*, 272-273; Denikin, *White Army*, 228; Mawdsley, *Russian Civil War*, 134; Lincoln, *Red Victory*, 260.
15. Denikin, *White Army*, 229; Whiting, *Soviet Air Power*, 6; Shumikhin, *Sovietskaia voennaia aviatsiia: 1917-1941*; 61; Neshkin and Shabanov, comp., *Aviatory*, "Rutkovskii, Viacheslav Stepanovich," 253-256.

16. Neshkin and Shabanov, comp., *Aviatory*, "Rutkovskii, Viacheslav Stepanovich," 253-256 and "Kul'tin, Leonid Aleksandrovich," 159-161.
17. Shumikhin, *Sovetskaia voennaia aviatsiia: 1917-1941*, 62; Kilmarx, *History of Soviet Air Power*, 54; Whiting, *Soviet Air Power*, 6.
18. Shumikhin, *Sovetskaia voennaia aviatsiia: 1917-1941*, 62; Stewart, *White Armies of Russia*; Lincoln, *Red Victory*, 261-263; Whiting, *Soviet Air Power*, 6.
19. Ulam, *Expansion and Coexistence*, 101-102; Lincoln, *Red Victory*, 262-263. For an excellent brief discussion of the shortcomings of the Kolchak military, see Mawdsley, *Russian Civil War*, 144-147; Denikin, *White Army*, 269-272.
20. Donaldson and Nogee, *Foreign Policy of Russia*, 52-53; Mawdsley, *Russian Civil War*, 232-233; Lincoln, *Red Victory*, 414.
21. Denikin, *White Army*, 238; Kilmarx, *History of Soviet Air Power*, 55; Taylor, *First World War*, 242; Mawdsley, *Russian Civil War*, 167; Vrangel', *Always With Honour*, 75.
22. Neshkin and Shabanov, comp., *Aviatory*, "Nadezhdin, Vadim Mikhailovich," 199-201, and "Rudnev, Evgenii Vladimirovich," 251-253.
23. Neshkin and Shabanov, comp., *Aviatory*, "Nadezhdin, Vadim Mikhailovich," 199-201; Lincoln, *Red Victory*, 214. While the Germans did provide military assistance to some anti-Soviet groups, Denikin notes that the Volunteer Army never cooperated with the former enemy. See Denikin, *White Army*, 134; Vrangel', *Always With Honour*, 69. Vrangel' called it the Volunteer Caucasian Army.
24. Jackson, *Red Falcons*, 14; Taylor, ed., *Combat Aircraft of the World*, 419-421; Green and Swanborough, *Complete Book of Fighters*, 535; James Streckfuss, "Sopwith Aircraft," in Boyne, ed., *Air Warfare*, vol.1, 574-575.
25. Jackson, *Red Falcons*, 13-14; Whiting, *Soviet Air Power*, 6-7; Green and Swanborough, *Complete Book of Fighters*, 223; Neshkin and Shabanov, comp., *Aviatory*, "Pankrat'ev, Aleksei Vasil'evich," 222-224, and "Chekhutov, Aleksandr Anikievich," 301-302.
26. Denikin, *White Army*, 222 and 213-216; Mawdsley, *Russian Civil War*, 170-172; Kilmarx, *History of Soviet Air Power*, 55; Jackson, *Red Falcons*, 14-15; Vrangel', *Always With Honour*, 81-83. In several locations, Denikin expressed his deep disappointment in the fact that the French and British military did not use their armed forces to assist the Whites. For example, see his autobiography *White Army*, 208.
27. Shumikhin, *Sovetskaia voennaia aviatsiia: 1917-1941*, 62; Neshkin and Shabanov, comp., *Aviatory*, "Stepanov, Ivan Petrovich," 270-272, "Kovan'ko, Aleksandr Aleksandrovich," 134-135, and "Tunoshenskii, Ivan Nikolaevich," 288-289.
28. Neshkin and Shabanov, comp., *Aviatory*: "Zverev, Fedor Trofimovich," 115, "Tkachev, Viacheslav Matveevich," 281-284, "Baranov, Viacheslav Grigor'evich," 28-29, and see also the military record of pilot "Strzhizhevskii, Vladimir Ivanovich," 273-275.
29. Kilmarx, *History of Soviet Air Power*, 55; Jackson, *Red Falcons*, 15-16; Mawdsley, *Russian Civil War*, 173-174; Denikin, *White Army*, 237-238; Vrangel', *Always With Honour*, 84-88.
30. Vrangel', *Always With Honour*, 89; Stewart, *White Armies of Russia*, 174-175; Denikin, *White Army*, 239; Mawdsley, *Russian Civil War*, 172; Whiting, *Soviet Air Power*, 7; Shumikhin, *Sovetskaia voennaia aviatsiia: 1917-1941*, 62; Neshkin and Shabanov, comp., *Aviatory*, "Bratoliubov, Georgii Aleksandrovich," 50-51, "Stroev, Mikhail Pavlovich," 275-277, "Bashko, Iosif Stanislavovich," 36-38, and "Romanov, Vladimir Aleksandrovich," 248-250.
31. Ibid; Mawdsley, *Russian Civil War*, 195-196; Denikin, *White Army*, 242; on 244 Denikin described the raid as both "brilliant and ineffective."

32. Denikin, *White Army*, 192-193 and 274-275; Lincoln, *Red Victory*, 199 and 294-295; Mawdsley, *Russian Civil War*, 197-199; Kilmarx, *History of Soviet Air Power*, 52-53; Stewart, *White Armies of Russia*, 209-230.
33. Denikin, *White Army*, 276; Lincoln, *Red Victory*, 296-298; Mawdsley, *Russian Civil War*, 200-201; Stewart, *White Armies of Russia*, 231-238.
34. Quote from Shumikhin, *Sovetskaia voennaia aviatsiia: 1917-1941*, 63; Mawdsley, *Russian Civil War*, 200-201. The quotation is likely a fabrication, although a White eyewitness to the execution may have survived and surrendered to the Seventh Red Army.
35. Mawdsley, *Russian Civil War*, 201-215; Kilmarx, *History of Soviet Air Power*, 55; Whiting, *Soviet Air Power*, 7; Denikin, *White Army*, 273-276 and 280-292; Vrangel', *Always With Honour*, 108-120.

CHAPTER TEN. SOVIET VICTORIES IN 1920 AND 1921

1. Denikin, *White Army*, 251-268; Libbey, *Russian-American Economic Relations*, 77.
2. Jackson, *Red Falcons*, 17-18; Stewart, *White Armies of Russia*, 361; Denikin, *White Army*, 293-306; Neshkin and Shabanov, comp., *Aviatory*, "Tkachev, Viacheslav Matveevich," 281-184; Mawdsley, *Russian Civil War*, 224; Vrangel', *Always With Honour*, 138-139.
3. Neshkin and Shabanov, comp., *Aviatory*, "Tkachev, Viacheslav Matveevich," 281-284; Denikin, *White Army*, 308; Lincoln, *Red Victory*, 216 and 436-437; Vrangel', *Always With Honour*, 152.
4. Denikin, *White Army*, 310; "Denikin, Anton Ivanovich (1872-1947)," in Michael T. Florinsky, ed., *Encyclopedia of Russia and the Soviet Union* (New York: McGraw-Hill Book Company, 1961), 130; Vrangel', *Always With Honour*, 140-146.
5. Quote from Denikin, *White Army*, 312, but see also 308; Vrangel', *Always With Honour*, 129-130.
6. Lincoln, *Red Victory*, 412-414; Vrangel', *Always With* Honour, 155 and 184-185; Denikin, *White Army*, 308; Kilmarx, *History of Soviet Air Power*, 56; Durkota, et al., *Imperial Russian Air Service*, "Aleksandr Mikhailovich Pishvanov," 95-97; Neshkin and Shabanov, comp., *Aviatory*: "Gartman, Maksimilian Evgen'evich," 71-73, "Zakharov, Anatolii Timofeevich," 114, "Zverev, Fedor Trofimovich," 115, "Kovan'ko, Aleksandr Aleksandrovich," 134-135, "Nadezhdin, Vadim Mikhailovich," 199-201, Strzhizhevskii, Vladimir Ivanovich," 273-275, and "Timofeev, Anatolii Konstantinovich," 280-281.
7. Neshkin and Shabanov, comp., *Aviatory*, "Naumov, Aleksandr Aleksandrovich," 203-204, "Modrakh, Sergei Karlovich," 191-192, and "Tunoshenskii, Ivan Nikolaevich," 288-289.
8. Vrangel', *Always With Honour*, 151; Stewart, *White Armies of Russia*, 370; Denikin, *White Army*, 13; Lincoln, *Red Victory*, 414.
9. Quote from Lincoln, *Red Victory*, 416; Denikin, *White Army*, 313; Mawdsley, *Russian Civil War*, 264-265; Vrangel', *Always With Honour*, 214-221.
10. Neshkin and Shabanov, comp., *Aviatory*, "Tkachev, Viacheslav Matveevich," 281-284; Lincoln, *Red Victory*, 416-417 and 436; Mawdsley, *Russian Civil War*, 262-268; Vrangel', *Always With Honour*, 219 and 228.
11. George Sanford, *Poland: The Conquest of History* (Amsterdam: Harwood Academic Publishers, 1999), 3; Hans Roos, *A History of Modern Poland* (New York: Alfred A. Knopf, 1966) 76-78 and see the map of Poland with the Curzon Line on p. 81; Mawdsley, *Russian Civil War*, 250-251.
12. Jerzy B. Cynk, *Polish Aircraft: 1893-1939* (London: Putnam & Company, 1971), 106-107.

13. Cynk, *Polish Aircraft: 1893-1939*, 107.
14. Vrangel', *Always With Honour*, 217; Lincoln, *Red Victory*, 408-416; Mawdsley, *Russian Civil War*, 252-253; Roos, *History of Modern Poland*, 78; Neshkin and Shabanov, comp., *Aviatory*, "Vologodtsev, Lev Konstantinovich," 64-65, and "Kul'tin, Leon Aleksandrovich," 159-161.
15. Shumikhin, *Sovetskaia voennaia aviatsiia: 1917-1941*, 64.
16. Roos, *History of Modern Poland*, 82-83 and 93-97; Mawdsley, *Russian Civil War*, 255-257; Donaldson and Nogee, *Foreign Policy of Russia*, 52-53; Vrangel', *Always With Honour*, 254-255.
17. Lincoln, *Red Victory*, 433-437; Vrangel', *Always With Honour*, 223 and 233; Jackson, *Red Falcons*, 20.
18. Shumikhin, *Sovetskaia voennaia aviatsiia: 1917-1941*, 65; Jackson, *Red Falcons*, 21; Durkota, et al., *Imperial Russian Air Service*, "Viacheslav Matveevich Tkachev," 281-284.
19. Vrangel', *Always With Honour*, 249-250; Lincoln, *Red Victory*, 437.
20. Vrangel', *Always With Honour*, 249; Stewart, *White Armies of Russia*, 372-373; Lincoln, *Red Victory*, 438-439; Mawdsley, *Russian Civil War*, 268; Neshkin and Shabanov, comp., *Aviatory*, "Kovan'ko, Aleksandr Aleksandrovich," 134-135.
21. Vrangel', *Always With Honour*, 259-265; Lincoln, *Red Victory*. 439; Mawdsley, *Russian Civil War*, 268.
22. Vrangel', *Always With Honour*, 278-279.
23. Vrangel' quoted from *Always With Honour*, 279.
24. Vrangel', *Always With Honour*, 283 and 287; Neshkin and Shabanov, comp., *Aviatory*, "Gartman, Maksimilian Evgen'evich," 71-73.
25. Lee, *Soviet Air Force*, 26; Raymond L. Garthoff, *Soviet Military Doctrine* (Glencoe, IL: The Free Press, 1953), 325; Shumikhin, *Sovetskaia voennaia aviatsiia: 1917-1941*, 65.
26. Neshkin and Shabanov, comp., *Aviatory*, "Matson, Garal'd Aleksandrovich," 185-187; Jackson, *Red Falcons*, 21; Kilmarx, *History of Soviet Air Power*, 56.
27. Kilmarx, *History of Soviet Air Power*, 56; Shumikhin, *Sovetskaia voennaia aviatsiia: 1917-1941*, 65-66.
28. Lincoln, *Red Victory*, 443-447; Mawdsley, *Russian Civil War*, 269-270.
29. Lincoln, *Red Victory*, 448-449; Stewart, *White Armies of Russia*, 376-377; Vrangel', *Always With Honour*, 313-327; Neshkin and Shabanov, comp., *Aviatory*, 28-29, 71-73, 114, 115, 134-135, 203-204, 251-253, and 281-284.
30. Kilmarx, *History of Soviet Air Power*, 58-60; Jackson, *Red Falcons*, 21-22; Evtuhov, et al., *History of Russia*, 618-619; Daniels, *Conscience of the Revolution*, 143-145 and 154.

CHAPTER ELEVEN. AIRCRAFT DEVELOPMENT, 1918-1924

1. Shumikhin, *Sovetskaia voennaia aviatsiia: 1917-1941*, 70; Morrow, *Great War in the Air*, 283; Jackson, *Red Falcons*, 22.
2. Glushko, ed.,"Zhukovskii, Nikolai Egorovich (1847-1921)," *Kosmonavtika entsiklopediia*, 114; Gregori Aleksandrovich Tokaev, *Betrayal of an Ideal* (London: Harvel Press, 1954), 187; Boyd, *Soviet Air Force*, 5; Lee, *Soviet Air Force*, 28.
3. G. S. Bushgens and S. L. Chernyshev, *TsAGI: Russia's Global Aerospace Research Center* (New York: Begell House Publishing, 2011), 20.

4. Boyd, *Soviet Air Force*, 5; Gunston, *Osprey Encyclopedia of Russian Aircraft*, 156–158 and 379–408. For a short biography of Tupolev and a nice description of his pre–World War II planes, see Pol Daffi and Andrei Kandalov, *A. N. Tupolev: Chelovek I evo samolety* (Moskva: Moskovskii rabochii, 1999), 11–12 and 37–101.
5. Bushgens and Chernyshev, *TsAGI*, 19–42; Whiting, *Soviet Air Power*, 8; Kilmarx, *History of Soviet Air Power*, 63.
6. Libbey, *Russian-American Economic Relations*, 77–78. Soviet currency continued to be aligned at an exchange rate of five rubles to one U.S. dollar until 1947.
7. Boyd, *Soviet Air Force*, 6–7; Gunston, *Osprey Encyclopedia of Russian Aircraft*, 286; Taylor, ed., *Combat Aircraft of the World*, 416.
8. Gunston, *Osprey Encyclopedia of Russian Aircraft*, 286; Taylor, ed., *Combat Aircraft of the World*, 603.
9. Gunston, *Osprey Encyclopedia of Russian Aircraft*, 286; Boyd, *Soviet Air Force*, 7.
10. Jackson, *Red Falcons*, 22–23; Gunston, *Osprey Encyclopedia of Russian Aircraft*, 12 and 87.
11. Gunston, *Osprey Encyclopedia of Russian Aircraft*, XIV, XXXI, and 88–95.
12. Green and Swanborough, *Complete Book of Fighters*, 471–478; Gunston, *Osprey Encyclopedia of Russian Aircraft*, 286–313; Taylor, ed., *Combat Aircraft of the World*, 594–603.
13. Gunston, *Osprey Encyclopedia of Russian Aircraft*, 99; Boyd, *Soviet Air Force*, 8; Taylor, ed., *Combat Aircraft of the World*, 311–312.
14. Boyd, *Soviet Air Force*, 8; Daffi and Kandalov, *A. N. Tupolev*, 15; Gunston, *Osprey Encyclopedia of Russian Aircraft*, 99–129; Taylor, ed., *Combat Aircraft of the World*, 311–312, 570–576; Whiting, *Soviet Air Power*, 10; V. P. Kozlov and V. G. Kazashvili, *Aviatsiia Rossii* (Monino: Muzei Voenno–Vozdushykh Sil, 2000), 8.
15. Gunston, *Osprey Encyclopedia of Russian Aircraft*, 87 and 286–287; Green and Swanborough, *Complete Book of Fighters*, 471; James Streckfuss, "Fokker Aircraft (Early Years, World War I)," in Boyne, ed., *Air Warfare*, vol. 1, 227–228; Christopher H. Sterling, "Fokker Aircraft (Post–World War I)," in Boyne, ed., *Air Warfare*, vol. 1, 228–229.
16. James Streckfuss, "Nieuport Aircraft," in Boyne, ed., *Air Warfare*, vol. 2, 451–452; Taylor, ed., *Combat Aircraft of the World*, 157–158; Green and Swanborough, *Complete Book of Fighters*, 224–225. Some sources credit Reinhold Platz for designing the D. VII, but his job involved smoothing out the wrinkles in production rather than designing aircraft. See James Streckfuss, "Platz, Reinhold (1886–1986)," in Boyne, ed., *Air Warfare*, vol., 497.
17. James Streckfuss, "Fokker Aircraft (Early Years, World War I)," in Boyne, ed., *Air Warfare*, vol. 1, 227–228; Taylor, ed., *Combat Aircraft of the World*, 157–158 and 272; Green and Swanborough, *Complete Book of Fighters*, 224–225 and 228–229.
18. Taylor, ed., *Combat Aircraft of the World*, 272–273; Donaldson and Nogee, *Foreign Policy of Russia*, 55; Gunter Schmitt, translated by Charles E. Scurrell, *Hugo Junkers and His Aircraft* (Berlin: VEB Verlag für Verkehrswesen, 1988), 130; Richard Byers, *Flying Man: Hugo Junkers and the Dream of Aviation* (College Station: Texas A&M University Press, 2016), 58.
19. John D. Hicks, *Republican Ascendancy: 1921–1933* (New York: Harper & Brothers Publishers, 1960), 51–52; James K. Libbey, *Alben Barkley: A Life in Politics* (Lexington: University Press of Kentucky, 2016), 148.
20. Libbey, *Russian-American Economic Relations*, 61 and 68.

21. Donaldson and Nogee, *Foreign Policy of Russia*, 55; Schmitt, *Hugo Junkers and His Aircraft*, 131–132; Byers, *Flying Man*, 54.
22. Armand Hammer with Neil Lyndon, *Hammer* (New York: Putnam Publishing Group, 1987), 115–118; Libbey, *Alexander Gumberg and Soviet-American Relations 1917-1933*, 142–143; Armand Hammer, *The Quest of the Romanoff Treasure* (New York: Paisley Press, 1932); Taylor, *Aerospace Chronology*, 72; Schmitt, *Hugo Junkers and His Aircraft*, 133; Byers, *Flying Man*, 25.
23. Schmitt, *Hugo Junkers and His Aircraft*, 132–133; Kilmarx, *History of Soviet Air Power*, 65–66; Boyd, *Soviet Air Force*, 9.
24. Edward L. Homze, *Arming the Luftwaffe: The Reich Air Ministry and the German Aircraft Industry 1919-39* (Lincoln: University of Nebraska Press, 1976), 2; Byers, *Flying Man*, 41.
25. Homze, *Arming the Luftwaffe*, 3; Byers, *Flying Man*, 70.
26. Donaldson and Nogee, *Foreign Policy of Russia*, 55; Leonard Schapiro, ed., *Soviet Treaty Series, 1917-1928*, vol. 1 (Washington, DC: Georgetown University Press, 1950), 381–382; Boyd, *Soviet Air Force*, 9; Homze, *Arming the Luftwaffe*, 9.
27. Schmitt, *Hugo Junkers and His Aircraft*, 135–136; Byers, *Flying Man*, 57.
28. Homze, *Arming the Luftwaffe*, 25; Green and Swanborough, *Complete Book of Fighters*, 312; Gunston, *Osprey Encyclopedia of Russian Aircraft*, XX and XXXI; Boyd, *Soviet Air Force*, 10; Kilmarx, *History of Soviet Air Power*, 65.
29. Boyd, *Soviet Air Force*, 13–14; Lee, *Soviet Air Force*, 29–30. For a more complete discussion see Scott W. Palmer, *Dictatorship of the Air: Aviation Culture and the Fate of Modern Russia* (New York: Cambridge University Press, 2006).

BIBLIOGRAPHY

Acton, Edward. *Russia: The Present and the Past*. London: Longman Group, 1986.
Alexander, Grand Duke of Russia. *Once a Grand Duke*. New York: Farrah & Rinehart, 1932.
Andreev, Igor'. *Boev'e samolety*. Moskva: Prostreks, 1994.
Barnhill, John. "Ely, Eugene (1886–1911)," in Boyne, ed., *Air Warfare*, vol. 1, 193–194.
Berg, A. Scott. *Lindbergh*. New York: G. P. Putnam's Sons, 1998.
Bilstein, Roger E. *Flight in America: From the Wrights to the Astronauts*. Baltimore: Johns Hopkins University Press, 1994.
Black, C. E., and E. C. Helmreich. *Twentieth Century Europe: A History*. New York: Alfred A. Knopf, 1963.
Blume, August G. "The Eastern Front: War Episodes of Russian Aviation." *Over the Front*, 5 (Winter 1990), 340–355.
Boyd, Alexander. *The Soviet Air Force Since 1918*. New York: Stein and Day, 1977.
Boyne, Walter J. *The Influence of Air Power upon History*. Gretna, LA: Pelican Publishing Company, Inc., 2003.
———. ed. *Air Warfare: An International Encyclopedia*, 2 vols. Santa Barbara, CA: ABC-CLIO, 2002.
Budiansky, Stephen. *Air Power: The Men, Machines, and Ideas that Revolutionized War, from Kitty Hawk to Iraq*. New York: Penguin Books, 2004.
Bunyan, James. *International Civil War and Communism in Russia, April–December 1918*. Baltimore: Johns Hopkins Press, 1936.
Bushgens, G. S., and S. L. Chernyshev. *TsAGI: Russia's Global Aerospace Research Center*. New York: Begell House Publishing, 2011.
Byers, Richard. *Flying Man: Hugo Junkers and the Dream of Aviation*. College Station: Texas A&M University Press, 2016.
Chant, Christopher. *Pioneers of Aviation*. New York: Barnes & Noble, 2001.
Churchill, Winston S. *The Unknown War: The Eastern Front*. New York: Charles Scribner's Sons, 1931.
Cochrane, Dorothy, Von Hardesty, and Russell Lee. *The Aviation Careers of Igor Sikorsky*. Seattle: University of Washington Press for the National Air & Space Museum, 1989.
Corum, James S., and Richard R. Muller. *The Luftwaffe's Way of War: German Air Force Doctrine 1911–1969*. Baltimore: Nautical & Aviation Publishing Company of America, 1998.
Creveld, Martin van. *The Age of Air Power*. New York: Public Affairs, 2011.
Cross, Wilbur. *Zeppelins of World War I*. New York: Barnes & Noble, 1991.

Crouch, Tom D. *The Bishop's Boys: A Life of Wilbur and Orville Wright*. New York: W. W. Norton & Company, 1989.

———. *Lighter Than Air: An Illustrated History of Balloons and Airships*. Baltimore: Johns Hopkins University Press, 2009.

———. *Wings: A History of Aviation from Kites to the Space Age*. New York: Smithsonian National Air and Space Museum in association with W. W. Norton & Company, 2003.

Cynk, Jerzy B. *Polish Aircraft: 1893–1939*. London: Putnam & Company, 1971.

Daffi, Pol, and Andrei Kandalov. *A. N. Tupolev: Chelovek i evo samolety*. Moskva: Moskovskii rabochii, 1999.

Daniels, Robert Vincent. *The Conscience of the Revolution: Communist Opposition in Soviet Russia*. New York: Simon and Schuster, 1969.

Delear, Frank J. *Igor Sikorsky: His Three Careers in Aviation*. New York: Dodd, Mead & Company, 1976.

Denikin, General A. I. *The White Army*. Cambridge, UK: Ian Faulkner Publishing, 1992.

Donaldson, Robert H., and Joseph L. Nogee. *The Foreign Policy of Russia: Changing Systems, Enduring Interests*. New York: M. E. Sharpe, 2002.

Durkota, Alan, et al. *The Imperial Russian Air Service: Famous Pilots and Aircraft of World War One*. Mountain View, CA: Flying Machines Press, 1996.

Duz', Petr D. *Istoriia vozdukhoplavaniia i aviatsii v Rossii. (Period do 1914 g.)*. 2nd ed. Moskva: Mashinostroenie, 1995.

Duz', P(etr) D. *Istoriia vozdukhoplavaniia i aviatsii v SSSR. Period pervoi mirovoi voiny (1914–1918 g.g.)*. Moskva: Gosudarstvennoe Nauchno-Tekhnicheskoe Izdatel'stvo Oborongiz, 1960.

Evtuhov, Catherine, et al. *A History of Russia: Peoples, Legends, Events, Forces*. Boston: Houghton Mifflin Company, 2004.

Farber, James. "Major de Seversky—Engineer." *Popular Aviation* 17 (August 1935), 87–88, 116, and 132.

Finne, K. N., edited by Carl J. Borrow and Von Hardesty. *Igor Sikorsky: The Russian Years*. Washington, DC: Smithsonian Institution Press, 1967.

Florinsky, Michael T., ed. *Encyclopedia of Russia and the Soviet Union*. New York: McGraw-Hill Book Company, 1961.

Franks, Norman. *Nieuport Aces of World War I*. Oxford, UK: Osprey Publishing, 2008.

Gaddis, Barbara Jackson. "American Economic Interests in Russia, August 1914–March 1917." M.A. thesis, University of Texas, 1966.

Gaddis, John Lewis. *Russia, the Soviet Union and the United States*. New York: McGraw-Hill, 1998.

Garthoff, Raymond L. *Soviet Military Doctrine*. Glencoe, IL: The Free Press, 1953.

Gilbert, Felix, with David Clay Large. *The End of the European Era, 1890 to the Present*. New York: W. W. Norton & Company, 1991.

Glushko, V. P., ed. "Zhukovskii, Nikolai Egorovich (1847–1921)," *Kosmonavtika entsiklopediia*. Moskva: Sovietskaiia entsiklopediia, 1985.

Golovine, Nicholas N. *The Russian Army in the World War*. New Haven: Yale University Press, 1931.

Graf, H(arold K.). *The Russian Navy in War and Revolution: From 1914 up to 1918*. Honolulu: University Press of the Pacific, 2002.

Grambow, Marion E., ed. *Who's Who in World Aviation and Astronautics*, vol. 2. Washington: American Aviation Publications, 1958.

Green, William and Gordon Swanborough. *The Complete Book of Fighters*. New York: Barnes & Noble, 1998.
Greger, René. *The Russian Fleet 1914–1917*. London: Jan Allan, Ltd., 1972.
Griffith, H. F. *RAF in Russia*. London: Hammond and Company, 1942.
Gunston, Bill. *The Development of Piston Aero Engines*. Newbury Park, CA: Haynes North America, 1999.
———. *The Osprey Encyclopedia of Russian Aircraft 1875-1995*. London, UK: Osprey Aerospace, 1995.
Hallion, Richard P. "Unlikely Partners: German-Soviet Aeronautical Cooperation, 1919–1933." American Institute of Aeronautics and Astronautics (January 2018), 1–21.
Hammer, Armand. *The Quest of the Romanoff Treasure*. New York: Paisley Press, 1932.
Hammer, Armand, with Neil Lyndon. *Hammer*. New York: Putnam Publishing Group, 1987.
Hardesty, Von. "Early Flight in Russia," in Higham, Greenwood, and Hardesty, eds. *Russian Aviation and Air Power in the Twentieth Century*. London: Frank Cass Publishers, 1998, 25–36.
Herring, George C. *From Colony to Superpower: U.S. Foreign Relations since 1776*. New York: Oxford University Press, 2008.
Hicks, John D. *Republican Ascendancy: 1921–1933*. New York: Harper & Brothers Publishers, 1960.
Higham, Robin, and Jacob W. Kipp, eds. *Soviet Aviation and Air Power: A Historical View*. Boulder, CO: Westview Press, 1977.
Higham, Robin, John T. Greenwood, and Von Hardesty, eds. *Russian Aviation and Air Power in the Twentieth Century*. London: Frank Cass Publishing, 1998.
Homze, Edward L. *Arming the Luftwaffe: The Reich Air Ministry and the German Aircraft Industry 1919–39*. Lincoln: University of Nebraska Press, 1976.
Ironside, W[illiam] E[dmund]. *Archangel, 1918–1919*. London: Constable & Company, 1953.
Jackson, Robert. *The Red Falcons: The Soviet Air Force in Action, 1919–1969*. London: Clifton Books, 1970.
Johnson, Herbert A. *Wingless Eagle: U.S. Army Aviation through World War I*. Chapel Hill: University of North Carolina Press, 2001.
Jones, David R. "The Beginnings of Russian Air Power, 1907–1922." In Higham and Kipp, eds. *Soviet Aviation and Air Power*. Boulder, CO: Westview Press, 1977, 15–33.
———. "The Birth of the Russian Air Weapon 1909–1914." *Aerospace Historian* 21 (Fall 1974), 169–171.
Karman, Theodore von. *Aerodynamics*. Ithaca: Cornell University Press, 1954.
Kennan, George F. *The Decision to Intervene*. New York: Atheneum, 1967.
———. *Russia Leaves the War*. New York: Atheneum, 1967.
Kennett, Lee. *The First Air War 1914–1918*. New York: The Free Press, 1991.
Kilmarx, Robert A. *A History of Soviet Air Power*. New York: Frederick A. Praeger, 1962.
———. "The Russian Imperial Air Forces of World War I." *Air Power Historian*, 10 (July 1963), 90–95.
Kipp, Jacob W. "Development of Naval Aviation, 1908–1975," in Higham and Kipp, eds. *Soviet Aviation and Air Power*, 140–141.
Kiuchariants, D[zhul'etta] A[rturovna]. *Gatchina: Khudozhestvennye pamiatniki*. Leningrad: Leninzdat, 1990.
Kort, Michael. *The Soviet Colossus: A History of the USSR*. Boston: Unwin Hyman, 1990.

Kozlov, V. P., and V. G. Kazashvili. *Aviatsiia Rossii*. Monino: Muzei Voenno-Vozdushykh Sil, 2000.

Kulikov, Viktor. "Aeroplanes of Lebedev's Factory." *Air Power History* 48 (Winter 2001), 4–17.

Lapchinskii, Aleksandr Nikolaevich. *Vozduzhnaia razvedka*. Moskva: Gosvoenizdat N.K.O., 1938.

Layman, R. D. *Naval Aviation in the First World War*. Annapolis: Naval Institute Press, 1996.

Lee, Asher. *The Soviet Air Force*. New York: John Day, 1962.

———. *The Soviet Air and Rocket Forces*. New York, Praeger, 1959.

Libbey, James K. *Alben Barkley: A Life in Politics*. Lexington: University Press of Kentucky, 2016.

———. *Alexander Gumberg and Soviet-American Relations 1917-1933*. Lexington: University Press of Kentucky, 1977.

———. *Alexander P. de Seversky and the Quest for Air Power*. Washington, DC: Potomac Books, an imprint of the University of Nebraska Press, 2013.

———. "Flight Training and American Aviation Pioneers." *American Aviation Historical Society Journal* 49 (Spring 2004), 28–36.

———. "The Making of a War Hero." *Aviation History* 20 (March 2010), 54–59.

——— *Russian-American Economic Relations 1763–1999*. Gulf Breeze, FL: Academic International Press, 1999.

———. "Sikorsky, Igor I. (1889–1972)" in Boyne, ed. *Air Warfare*, 567.

Lieven, Dominic. *The End of Tsarist Russia*. New York: Viking, 2015.

Lincoln, W. Bruce. *Passage Through Armageddon: The Russians in War and Revolution*. New York: Oxford University Press, 1994.

———. *Red Victory: A History of the Russian Civil War*. New York: Da Capo Press, 1999.

———. *The Romanovs: Autocrats of All the Russias*. Garden City, NY: Anchor Books, 1981.

Lovett, Christopher C. "Russian and Soviet Naval Aviation," in Higham, Greenwood, and Hardesty, eds., *Russian Aviation and Air Power in the Twentieth Century*, 108–125.

"Maj. Alexander P. de Seversky," in Biographies of Pioneer Aviators, Orvil A. Anderson Collection, U.S. Air Force Historical Research Agency, 168.7006-47, p. 2.

Mawdsley, Evan. *The Russian Civil War*. New York: Pegasus Books, 2005.

Menning, Bruce W. *Bayonets Before Bullets: The Imperial Russian Army, 1861–1914*. Bloomington: Indiana University Press, 1992.

Mikheyev, Vadim. *Sikorsky S-16*. Stratford, CT: Flying Machines Press, 1997.

Miller, Russell. *The Soviet Air Force at War*. Alexandria, VA: Time-Life Books, 1985.

Montefiore, Simon Sebag. *The Romanovs 1613–1918*. New York: Alfred A. Knopf, 2016.

Moolman, Valerie. *The Road to Kitty Hawk*. Alexandria, VA: Time-Life Books, 1980.

Morris, Edmund. *Theodore Rex*. New York: Random House, 2001.

Morrow, John H. Jr. *The Great War in the Air: Military Aviation from 1909 to 1921*. Washington, DC: Smithsonian Institution Press, 1993.

Nekrasov, George. *North of Gallipoli*. Boulder, CO: East European Monographs, 1992.

Neshkin, M. S., and V. M. Shabanov, comps. *Aviatory-kavaleri ordena Sv. Georgiia iGeorgievskovo oruzhiia perioda Pervoi mirovoi voiny 1914–1918 godov*. Moskva: Rosspen, 2006.

Nowarra, Heinz J., and G. R. Duval. *Russian Civil and Military Aircraft 1884–1969*. London: Fountain Press, 1971.

Opdycke, Leonard E. *French Aeroplanes Before the Great War*. Atglen, PA: Schiffer Publishing, 1999.

Palmer, Scott W. *Dictatorship of the Air: Aviation Culture and the Fate of Modern Russia.* New York: Cambridge University Press, 2006.
Phillips, Hugh D. *Between the Revolution and the West: A Political Biography of Maxim M. Litvinov.* Boulder, CO: Westview Press, 1992.
"Report of the November Meeting." *The Adventurer* (December 1940), 8.
Riaboff, Alexander. *Gatchina Days: Reminiscences of a Russian Pilot.* Washington, DC: The Smithsonian Institution Press, 1986.
Robertson, Bruce. *Air Aces of the 1914–1918 War.* Fallbrook, CA: Aero Publishers, 1964.
Roos, Hans. *A History of Modern Poland.* New York: Alfred A. Knopf, 1966.
"Russian Flyer with One Leg Seeks U.S. Job." *San Francisco Examiner*, April 22, 1918.
Sanford, George. *Poland: The Conquest of History.* Amsterdam, Netherlands: Harwood Academic Publishers, 1999.
Sanger, Ray. *Nieuport Aircraft of World War One.* Wiltshire, UK: Crowood Press, 2002.
Saul, Norman E. *Sailors in Revolt: The Russian Baltic Fleet in 1917.* Lawrence: The Regents Press of Kansas, 1978.
Saul, Norman E. *War and Revolution: The United States and Russia, 1914–1921.* Lawrence: University Press of Kansas, 2001.
Schapiro, Leonard, ed. *Soviet Treaty Series, 1917–1928,* vol. 1. Washington, DC: Georgetown University Press, 1950.
Schatunowski, George. "The Civil War to the Second World War," in Asher Lee, ed., *The Soviet Air and Rocket Forces.* New York: Frederick A. Praeger, 1959.
Schmitt, Gunter, translated by Charles E. Scurrell. *Hugo Junkers and His Aircraft.* Berlin: VEB Verlag für Verkehrswesen, 1988.
Sergievsky, Boris. *Airplanes, Women, and Song: Memoirs of a Fighter Ace, Test Pilot, and Adventurer.* Syracuse: Syracuse University Press, 1999.
Seton-Watson, Hugh. *Eastern Europe Between the Wars 1918–1941.* Boulder, CO: Westview Press, 1986.
Seversky, Alexander P. de. *Air Power: Key to Survival.* New York: Simon & Schuster, 1950.
———. "I Owe My Career to Losing a Leg." *Ladies Home Journal* 61 (May 1944), 20–23 and 164–169.
Shavrov, V. B. *Istoriia konstruktsii samoletov v SSSR do 1938g.* Moskva: Mashinostroenie, 1986.
Shumikhin, V. S. *Sovetskaia voennaia aviatsiia: 1917–1941.* Moskva: Nauka, 1986.
Sikorsky, Igor I. *The Story of the Winged-S*, 3rd ed. New York: Dodd Mead & Company, 1944.
Simakov, Boris. "Russkaia aviatsiia v pervoi mirovi voine." *Vestnik vozdushnovo flota* 5 (1952), 89.
Sorokin, A. K., otv. redactor. *Rossiia v Pervoi mirovoi voine 1914–1918.* Moskva: Politicheskaia entsiklopediia, 2014.
Sterling, Christopher H. "Fairey Aircraft," in Boyne, ed., *Air Warfare*, vol. 1, 213–214.
Stewart, George. *The White Armies of Russia: A Chronicle of Counter-Revolution and Allied Intervention.* New York: Russell & Russell, 1970.
Stokesbury, James L. *A Short History of Air Power.* New York: William Morrow and Company, 1986.
Stone, Norman. *The Eastern Front 1914–1917.* London, UK: Penguin Books, 1998.
Strakhovsky, Leonid I. *Intervention at Archangel.* Princeton: Princeton University Press, 1944.

Streckfuss, James. "Caudron Aircraft (Early Years)," in Boyne, ed. *Air Warfare*, vol. 1, 121.
———. "De Havilland Aircraft (Early Years and World War I)," in Boyne, ed. *Air Warfare*, vol. 1, 167.
———. "Royal Aircraft Factory," in Boyne, ed. *Air Warfare*, vol. 2, 534–535.
———. "Sopwith Aircraft," in Boyne, ed. *Air Warfare*, vol. 2, 574–575.
Taylor, A. J. P. *The First World War*. New York: G. P. Putnam's Sons, 1980.
Taylor, John W. R., ed. *Combat Aircraft of the World from 1909 to the Present*. New York: G. P. Putnam's Sons, 1969.
———. *The Lore of Flight*. New York: Barnes & Noble, 1996.
Taylor, Michael A. J. *The Aerospace Chronology*. London: Tri-Service Press, 1989.
Taylor, Swanee. "Seversky: An Ace's Place Is in the Air." *New York World Telegram Metropolitan Weekend Magazine* (December 12, 1936), 16.
Tokaev, Gregori Aleksandrovich. *Betrayal of an Ideal*. London: The Harvel Press, 1954.
Trunov, Konstantin I. *Petr Nesterov*. Moskva: Sovetskaiia Rossiia, 1975.
Tuchman, Barbara W. *The Guns of August*. New York: Bantam Books, 1989.
Ulam, Adam B. *Expansion and Coexistence: The History of Soviet Foreign Policy 1917–67*. New York: Frederick A. Praeger, 1968.
Vitarbo, Gregory. *Army of the Sky: Russian Military Aviation before the Great War, 1904–1914*. New York: Peter Lang Publishing, 2012.
Vrangel', General Petr N. *Always With Honour*. New York: Robert Speller & Sons, 1957.
Wade, Rex A. *The Russian Revolution, 1917*. New York: Cambridge University Press, 2005.
White, D. Fedotoff. *The Growth of the Red Army*. Princeton: Princeton University Press, 1944.
White, John A. *The Siberian Intervention*. Princeton: Princeton University Press, 1950.
Whiting, Kenneth R. *Soviet Air Power*. Boulder, CO: Westview Press, 1986.
Woodling, Bob, and Taras Chayka. *The Curtiss Hydroaeroplane: The U.S. Navy's First Airplane, 1911–1916*. Atglen, PA: Schiffer Publishing, Ltd, 2011.
Young, Warren R. *The Helicopters*. Alexandria, VA: Time-Life Books, 1982.

INDEX

Page numbers in *italics* indicate maps.

Ader, Clément, 3
aerodynamics, 16
Aeronautical Training Park, Volkhov Field, 4, 5, 6, 22
Aeronautics Unit (Section), 26, 28, 29–30, 34, 46
aircraft, imported: obsolescence of, 130; previous year models, 13–14, 193; for Soviet squadrons, 143–44, 168; for Whites, 156. *See also specific types of aircraft*
Aircraft Company, Ltd. (Airco), 140
aircraft production: early Soviet era, 130–31; in January 1916 versus August 1914, 90–91; by Poles at Mokotov Aerodrome, Warsaw, 167; in Russia, for Great War, 7, 10–14; Russian, during Great War, 194; Russian Revolution (1917) and, 63–64; Soviets nationalize factories for, 133. *See also* Anatra Aircraft Company; Dukh Company; Grigorovich, Dimitrii P.; Lebedev Aeronautics Company; Odessa Aero Club; Russo-Baltic [Railroad] Wagon Company; Shchetinin Works
aircraft repairs and support personnel: numerous aircraft types/models and, 46, 194; for Poland at Mokotov Aerodrome in Warsaw, 166–67; skills needed for, 45–46; train cars and, 133, 145, 183
Akashev, Konstantin V., 130

Albania, Balkan League and, 26
Albatros planes, 163; C.Ia, 95, 97; D. 1 and D. 2, 144; W.4, 111
Albion, Operation (German), 120
Alekhnovich, Georgii V., 85–86
Aleksandr, 103, 106
Aleksandr Mikhailovich, Grand Duke: air operations under, 91, 193; aircraft purchases and, 24; Blériot's flight and, 9–10, 192; escape from Soviet Russia, 200n17; fired as Russia's military aviation leader, 61; flight training and, 18, 19, 23; on foreign planes, 51–52; on Gatchina Palace air show, 16; on Il'ia Muromets, 58; military air arm inspectorate and, 35; problems of multiple machines and, 46; Sevastopol' Aviation School and, 20–21; on Shidlovskii's promotion and the EVK, 59; Tkachev replaces, 41
Aleksandra Feodorovna, 92
Alekseev, Mikhail V., 84, 90, 129
Alexis, Bishop (Russian Orthodox Church), 21
Allied Powers: on Brest-Litovsk Treaty and Russia fighting Germans, 139; on Decree on Peace, 128; on German aircraft and construction, 188–89; German and Russian reparations owed to, 185–86; Great War and, 1; Kolchak recognized as White leader by, 146; war materiel for Kolchak, 148. *See also* France; Great Britain; Italy; Russia; United States
All-Russian Aero Club, 124

All-Russian Board for the Administration of the Air Fleet, 130
All-Russian Cooperative Society, Ltd. (Arcos), 162, 179–80
Almaz, 65, 77, 79, 103
Amalia (Turkey), 77
amphibious operations, Black Sea Fleet and, 103–4
Anatra Aircraft Company, 12, 13, 90, 148
Angern, Lake, Germans encroaching in the Baltic and, 96–98, 99
ANT-4 aircraft, 179, 197
"Antoinette" aircraft engine, 9
Antonov-Ovseenko, Vladimir, 129
Anzani, Alessandro, 9
Anzani motors, 9, 11, 12, 179
Apsheron, 78
Arensburg naval air station, 73, 119–20
Argeev, Pavel V., 114–15
Argus engines, 14, 28, 53, 99
Arkhangel'sk: Allied troops and planes land in, 141; aviation engines imported through, 91; British and U.S. embassies' escape route to, 139; foreign commerce during Great War through, 48; Soviet military closing land connection with Murmansk, 147. *See also* North Russia; Slavic-British Aviation Corps
Arkhangel'skii, Aleksandr A., 133
armed forces, Russian, Soviet Order No. 1 disruption of, 109
Armed Forces of South Russia, 152, 153–54, 161, 169. *See also* Denikin, Anton I.; Vrangel', Petr Nikolaevich
Artseulov, Konstantin K., 182
Astra (dirigible), 8, 85
Atlantic Ocean, Lindbergh's flight across, 9
Augsburg, 64
Austria (Austro-Hungarian Empire): Brusilov's attack on, 100; Czech and Slovak soldiers against, 138; Fourth Army, 67, 79; Great War and, 1; Russians' attack on, 38; Russia's Fourth Army meets five divisions of, 42; Russia's Southwestern Front and, 66–67; Serbian nationalist kills heir to throne of, 55
Aviation Board, 130–32, 133. *See also* Red Air Force

Aviation Guarantee Committee, 189
Avksentiev, Nikolai D., 148
Avro 504K, British, copied as Soviet U-1 training plane, 183

Baian (cruiser), 122
Bakhirev, Mikhail K., 121–22
Bakhvalov, Aleksandr, 160, 217n34
Balkan League, 26–28
Balkan Wars, First and Second, 27–28
balloons, manned: artillery observation at Northwestern Front, 78; Baltic and Black Sea operations, 103; Great War and, 6–7; invention of, 4–5; Nizhevskii and, 85; Rudnev and, 22; Russia's military use of, 192; Russo-Japanese War and, 4, 5–6
Baltic Fleet: aircraft, 34, 51, 72; aircraft in Black Sea Fleet versus, 77; Central Committee and Congress, 111; Russo-Japanese War and, 3; Seversky and, 98. *See also* Essen, Nikolai Otto von; Kanin, Vasilii A.
Baltic Sea region: balloon operations, 103; blockades on aircraft imports to Russia, 47–48; German seaplane shot down in, 95; Germans in, Russian Empire and, 206n24; Germans planning Russian invasion along, 71–72; during the Great War, 33; iced over in winter, 107; Russian air stations and aircraft in, 73; Russian mine offensive in, 37, 64–65; Russian naval and aviation personnel in, 79
Baranov, Viacheslav G., 156–57, 175
Baranovichi, 35, 84–85, 94
Bashko, Iosif S., 79–81, 132–33, 158, 208n13
Beaver (torpedo boat), 26
Belgium, Schlieffen Plan and invasion of, 36–37
Bell, Alexander Graham, 3
Belousovich, Nikolai I., 132, 143, 147, 148
Beneš, Eduard, 138
Bereznik Aerodrome, 140, 142, 143, 146
Birmingham, 19
Black Sea: balloon operations, 103; Bulgaria and Turko-German position in, 79; Bulgaria on Russia during

Great War in, 28; German and Austrian occupation of, 131; during the Great War, *31*; international trade blocked in, 48–49; minelaying around Bosporus and, 105; opening of, after November 1918, 128, 153; Revolution, Order No. 1 and control of, 194–95; Russian aircraft in Baltic Sea Fleet versus, 77; Russian naval and aviation personnel in, 79; Russian task force in, 65; spring 1917 and naval officer/sailor relationship in, 111

Black Sea Fleet: amphibious operations and, 103–4; Caucasus campaign against the Turks by, 104–5; Communication Service, Air Force detachment, hydroplanes and, 29, 34; against German and Austrian troops, 103; ships and planes of, 112; winter operations, 106

Blériot, M. Louis, and Model XI monoplane, 9, 10, 11, 12, 17, 192

Blinov, Sergeant, 75, 76

BMW motors, 184, 190

Bogatyr, 37

Boleslaw I, King, 166

Bolshevik Party: Congress of Soviets and, 113; Constituent Assembly and, 111; on end to the war, 117; Lenin on new government formation and, 126, 128; Provisional Government and, 108; on tsarist pilots, 131. *See also* Lenin

Boruna, EVK attack on German reserve division near, 99

Bosporus, 77, 105. *See also* Black Sea

Brandenburg plane, Austrian, 102

Bratoliubov, Georgii A., 157

Breslau, 49, 50, 51, 103, 105, 112

Brest-Litovsk, as Il'ia Muromets base, 81–82

Brest-Litovsk, Treaty of, 126, 127, 131, 138, 195

Britain. *See* Great Britain

Brusilov, Aleksei A., 46, 66–67, 100, 102–3, 107

BSh-2 (*Bronirovanni Shturmovik* No. 2), 183–84

Budennyi, Semen M., 157, 167

Bulgaria, 26–28, 78, 172–73

Caudron, Gaston and René, G. 4 biplanes of, 141

Central Aero-Hydrodynamic Institute (TsAGI), 4, 133, 178, 179, 197

Central Aviation Park, Moscow, 133

Central Committee of the Baltic Fleet, 111

Central Executive Committee, Congress of Soviets and, 113

Central Powers: Bulgaria joining with, 28, 79; new Soviet government on cease-fire with, 117; Seversky and Soviet government peace treaty with, 213n5; support for Whites by, 154; Treaty of Brest-Litovsk and, 126; Ukrainian peace treaty with, 137–38. *See also* Austria; Germany; Turkey

Centralne Warsztaty Lotnicze (Central Aviation Workshops), Warsaw, 167

Chaikovskii, Nikolai V., 141

Chapaev, Vasilii I., 151–52

Chaplygin, Sergei Alekseevich, 179

Charles, Jacques, 5

Chekhutov, Aleksandr A., 110, 155

Chernov, Viktor, 128

Chief Directorate of Nationalized Aircraft Factories (*Glavkoavia*), 180

Chukhovskii, Vasilii G., 174

Clerget engine, 75, 76

coal, 65, 67

Collishaw, Raymond, 155

Commissariat of Foreign Trade, 179

Commission for the Planning and Construction of Piloted Aerostats, Russia, 8

Committee for Strengthening the Air Fleet, Russia, 10, 20

Committee on Aviation, 130

Communist Party: OVDF and, 190–91. *See also* Bolshevik Party

Congress of Soviets, 113; Fourth, 138; Second, 126–27

Constituent Assembly, 113–14, 128

Cossacks: Armed Forces of South Russia and, 154, 156; International Women's Day protests (1917) and, 108; in Kuban Cossack region, 129–30, 170–71; Siberian Army, 149; soldiers, in Black Sea, 104; White Army in South Russia and, 161. *See also*

Gorshkov, Georgii Georgievich;
  Tkachev, Viacheslav M.
Council of People's Commissars, 123
Crimea: during Russian Civil War, *136*;
  Soviet conquest of, 173–74, 196;
  Whites in, 162, 163, 164, 165, 166
Croix, Félix du Temple de la, 3
Crouch, Tom D., 3
Curtiss, Glenn A., 18
Curtiss Aeroplane and Motor Company,
  2, 14, 48
Curtiss Hudson Flier, 19
Curtiss seaplanes and flying boats, 77
Curzon, George N., 153
Curzon Line, 153, 166, 169
Czech Corps, 138–39, 141
Czech Legion, 138, 148–49
Czechoslovakia, 138, 148

D. VII pursuit plane, 184–85, 219n16
D. XI single-seat pursuit biplane, 185
D. XIII plane, 185
Dägo Island (Baltic), 64, 74, 120
de Havilland, Geoffrey, 14, 140
Decree on Land, 127
Decree on Peace, 127, 128, 129
Deka Company of Aleksandrov, 8
Denikin, Anton I.: military arms and
  equipment for, 153–55; "Moscow
  Directive" by, 157; moves troops to
  Soviet heartland, 152; resignation of,
  163; Soviet opposition by, 129;
  Volunteer Army and, 130; on White
  Army, 161, 216n26
Deperdussin (aircraft company), 11, 29
desertions, by peasants and conscripted
  soldiers, 109–10
D.H. 4s (single-engine biplane
  reconnaissance-bombers), 14,
  140, 148, 180
D.H. 9s, 140, 143–44, 154–55, 157, 169, 170
Dinamika polëta (*The Dynamics of Flight*)
  (Zhukovskii), 17
Directorate of the Military Aerial Fleet,
  91, 92, 113, 193
dirigibles, 7–9
Diterikhs, Lieutenant, 96–98
Dmitriev, Radko, 68, 82

Dnepr, 29
Dragomirov, Mikhail, 163
Dudorov, Boris, 118, 119
Dukh Company: aircraft production by, 7,
  11, 13, 45, 90, 200n12; civil war and
  nationalizing of, 144; FBA flying boats
  license-built by, 77; machine gun
  synchronization and, 63–64; military
  airplane competitions and, 29; navy
  contract for Nieuports, 112; renamed
  GAZ-1, 144–45; revolution and, 118
Duma, 52, 54, 108, 109
Dybovskii, Viktor Vladimirovich, 24

East Prussia, Russian army invasion of, 38,
  43, 44
Eberhardt, Andrei A., 20, 36, 49, 50–51,
  103–4, 105
Edinçik (Turkish steamer), 77
Efimov, Mikhail Nikolaevich, 21, 24
Egorov, Aleksandr I., 156, 167
Eindecker monoplane, 62–63, 74
Ekaterina (Catherine) the Great,
  Empress, 17
Eleventh Corps Squadron, 39–40, 41
Elpidifor, 104
Ely, Eugene, 19
engines, aircraft (motors): air reconnais-
  sance and limitations on, 45; for D.H.
  9s, 140, 144; early, 9; GAZ-2 and
  GAZ-4 as powerplants for, 190;
  imported, 9, 11, 14, 15, 46, 47, 94, 193;
  improvements to, 46; Renault, 16;
  Rudnev and problems with, 22;
  Russia's production limitations for,
  193; shortages (1916), 91; U.S. versus
  Russian during Great War, 14–15
England. *See* Great Britain
English Channel, Blériot's flight across, 9
Essen, Nikolai Otto von, 35–36, 37, 64, 73,
  74, 206n24
Estonia, 158–59, 160
EVK (Eskadra Vozdushnikh Korablei):
  developing anarchy and headquarters
  of, 117; flight-testing planes, 59–60;
  Grand Duke Aleksandr and, 92;
  headquarters relocation, 84; Imperial
  Russian Army support by, 79;

machine guns for, 63; Nizhevskii and, 85–86; pilot promotions in, 91; practice sessions for, 60–61; revolution's effect on, 113; Romanovskii's praise for, 82; Shidlovskii and formation of, 59; spring 1917 plans for, 107

Fairey Aircraft Company, 144
Falkenhayn, Erich von, 38, 78
Farman aircraft: importing difficulties due to war of, 48; Model III biplane, 11; Model IV primary trainers, 12, 17–18, 26; Model VII, Balkan War and, 27; as Russian license-built aircraft, 29
FBAs. *See* Franco-British Aviation flying boats
Ferdinand, King of Bulgaria, 78
Fiat motor, 181
fighter aircraft, 61–63, 100
Finland, 124, 126, 160
Finland, Gulf of, violence in, 110–11, 211n13
Finne, Konstantin N., 59
First Russian Aerostatics Company, 11
flight training: aerodromes for, 23; for Great War, 16–18; for Imperial Russian Army and Navy, 10; in land-based pursuit aircraft, 111; Rudnev and, 22; Russian deficiencies in, 193; for sea officers, 19; Soviet facilities for, 196. *See also* Imperial All-Russian Aero Club; Sevastopol' Aviation School
Fokker, Anthony H. G., 62, 155, 184–85
Fokker Dr I triplane, 155, 157
Fokker Eindecker, 74
fortresses: along western portion of Russian Empire, 6; aviation squadrons and, 34, 35; Novogeorgievsk, 22; Russo-Japanese War (1904–1905) and, 5; Sukhomlinov on abolishing, 25
Fourth Army, 42, 115, 116
France: aircraft engines exported to Russia by, 15, 91; aircraft inventory (1916), 89; aircraft production during Great War, 194; aircraft production in Russia versus, 14; Aleksandr's aircraft purchases from, 24; debts from Great War, 186; Escadrilles, Russian air squadrons compared with, 34; fighter aircraft development and, 61–62; Great War and, 1; military mission moves to Moscow, 139; Polish-Soviet war (1920) and, 169; Revolutionary, first military air unit of (1794), 5; Russian aircraft manufacturing and, 13, 29; Russian orders during Great War from, 48, 68; Russian pilots in North Russia and, 139–40; Russia's military flight development and, 192, 193; Schlieffen Plan for war against, 36–37; Vrangel' seeking Russian Army support from, 172–73; Whites' evacuation from Crimea and, 175. *See also* Deperdussin; Morane-Saulniers aircraft; Nieuports; Voisins
Franco-British Aviation (FBA) flying boats: in the Baltic Sea, 34, 72, 73; in the Black Sea, 77; bombing mission against Germans, 76–77; Essen's request for, 64; from Ösel Island, 73–74
Franz Ferdinand, archduke, assassination, 1
Free-Romanian Air Squadron, 115
French Colonial Division, Ukraine, 155
Friederich Carl, 64
Friederich der Grosse, 122
Friedrichstadt railway station, Russians bombing Germans at, 85–86, 93–95
Frunze, Mikhail V., 150, 152, 174–75
Fundamental Laws of 1832, tsar's revisions to (1905), 109

Gaida, Rudolf, 150
Gallipoli, 78
Garros, Roland, 61–62
Gartman, Maksimilian E., 163, 172, 175
Gastambide, Antoinette, 9
Gatchina Palace flight training center: early Soviet era program, 131; Officers' Aeronautics School at, 20, 22–23; program at, 16–18, 23; Sevastopol' pilot candidates and, 26
GAZ-1 (*Gosudarstvennyi Aviatsionnyi Zavod*-1, State Aviation Factory No. 1),

144–45, 180, 181–82; British Avro
504K at, 183
GAZ-2, as motor workshop, 190
GAZ-3 (Petrograd/Leningrad), 182
GAZ-4 (radial powerplant), 190
GAZ-5 (Moscow), 183
GAZ-10, in Taganrog, 181
GAZ-23 (Leningrad), 183
Gazelle, 64
Genoa Trade Conference (Italy), 185, 186
German Eighth Army, 43–44, 120
German Eleventh Army, 67, 68–69, 71–72, 79
German First Army, 36
German Fliegerabteilung, 34
German Ninth Army, 59
German Twelfth Army, 59
Germany: aircraft built with Russians, 190; aircraft inventory (1916), 89; closing Black Sea access in Great War, 48–49; dirigible development in, 8; Garros shot down by, 62; Great War and, 1; military assistance to anti-Soviet groups by, 216n23; Reparations Commission bill to, 186; Russian aircraft versus aircraft of, 2; Soviet Russia's collaboration on military aviation with, 185, 189–90; Treaty of Rapallo with Soviet Russia and, 187, 189, 197–98; Ukraine occupation by, 137–38; unopposed offensive into Russia (1918), 117. *See also* U-boats
gidroplan (hydroplanes), 2
Giffard, Jules-Henri, 7
Gil, Stepan, 128
Gil'sher, Georgii U. V., 114–15
gliders, 9, 14, 91
Gnome engines, 13, 15, 17–18, 71, 121, 190
Godzisk Aerodrome, 70
Goeben (Turko-German battlecruiser), 49, 50, 51, 77, 103
Golovine, Nicholas N., 102
Gorshkov, Georgii Georgievich, 60, 61, 80, 81
Grand (four-engine aircraft), 12, 14–15, 53
Grazhdanin (dreadnought), 122
Great Britain: aircraft engines exported to Russia (1916) by, 91; Allied troops and planes in North Russia and, 141–42; anger over Decree on Peace by, 128–29; debts from Great War, 186; declares war on Germany (1914), 37; disbands RAF squadrons supporting Whites, 161; Great War and, 1; Kolchak recognized as White leader by, 146; military arms and equipment for Denikin (1919), 153–55; military mission moves to Moscow, 139; Russian aircraft manufacturing and, 13; Russian embassy moves to Vologda, 139; Russian orders during Great War for planes and motors from, 48; on Russian pilots in North Russia, 139–40; Serbian army remnants and, 79; Vrangel' seeking Russian Army support from, 172–73; White support in Crimea by, 163; Whites' evacuation from Crimea and, 175; withdraws from Crimea, 166
Great Retreat, 32; aircraft and, 69–70, 71; flight during, 73–88. *See also* Baltic Sea region; Essen, Nikolai Otto von
Great War: aircraft inventory and expectations for shortness of, 47; D. VII aircraft provisions in armistice for, 184; fall 1914 campaign, 41–57; height of air war (1916), 89–105; *Magdeburg* capture and, 37–38; new aircraft roles (1915) for, 58–72; 1917 Revolution impacting squadrons for, 106–22; Russian versus German aircraft during, 2; Russians invade East Prussia and, 38; Schlieffen Plan for, 36–37; start of, 1; U.S. aircraft before, 4; Western versus Eastern Front of, 6–7, 67, 89–90
Greece, Balkan League and, 26–27
Grezo, Petr P., 70
Grigorovich, Dimitrii P.: aircraft availability (1917), 130–31; Essen's request for flying boats by, 64; flying boats, built by Shchetinin, 77; flying boats, White air squadrons (1920) and, 163; as indigenous aircraft designer, 193; Istrebitel' No. 2 development and, 181–82, 197; limitations of indigenous aircraft of, 184; M-5 flying boat with machine

INDEX

gun, 100; M-11 flying boat fighter, 111; *morskoi* (naval) aircraft and, 11–12; *morskoi* aircraft and, 14; TsAGI and, 133
Grinshteii, Lev K., 137
Guchkov, Aleksandr I., 60

Hackel, Jacob M., 28–29
Hague Conference (1899), 5
Hamidiye (Turkish cruiser), 49
Hammer, Armand, 187
Hannover-Roland CL IIa aircraft, 167
Hansa und Brandenburgische Flugzeug-Werke, 102
Hasse, Otto, 189
Heinkel, Ernst, 102
helicopters, Sikorsky and, 12
Herring, Augustus M., 3
Hindenburg, Paul von, 43–44
Hispano-Suiza engine, 107, 185, 190
Hoffmann, Max von, 43
hydroplanes, 2, 29, 34

Ianchenko, Vasili I., 100–101, 114–15
Iankovius, Viktor F., 94–95
Ianushkevich, Nikolai, 84
Il'ia Muromets (reconnaissance-bomber): Argus engines for, 14–15; armaments on board, 61; availability (1917), 130–31; Bashko's damaged, 81–82; bombing German train at Przevorsk (Galicia), 79–80, 208n13; combat missions during Great War, 92–95; construction and testing of, 54–55; at Eastern and Romanian fronts (summer 1917), 113; German retaliatory attack against, 80–81; Il-2, 184; Il-400 and Il-400B, 182; Kievskii models, 83, 99; Northwestern Front and, 58; production constraints, 194; production ended, 118; record of small single-engine planes versus, 86; Red Air Force flies last of, 174; replacement of, 178–79, 197; Rudnev training pilots for, 22; in Russian Air Force Museum, 211–12n22; Russian crew burning of, to keep from Germans, 117; Russo-Baltic building of, 12; Russo-Baltic engines for, 15–16; Sikorsky S-16 and, 13; Sikorsky's modifications and redesigns of, 55–56, 58–59, 83–84; special aviation group of Red Air Force and, 158; with Sunbeam engines at Stan'kovo Aerodrome, 99
Il'yushin, Sergei V., 183–84, 197
Imperial All-Russian Aero Club, 9–13, 16, 19, 87
Imperial Russian Army: aerial reconnaissance and, 193; aeronauts' training park, 4–5; aviation squadrons and, 34–35; aviation technical school and flight-training of, 10; Bulgarian army invasing Serbia and, 79; Caucasus campaign against the Turks by, 104; Central Powers' attack response by, 66; demobilization (1918), 123; dirigibles of, 8; flight training for Navy versus, 18–19; Great War aircraft of, 2, 19; military airplane competitions and, 28–29; remnants (1917), 116–17; Stavka and, 50
Imperial Russian Navy: aerial reconnaissance and, 193; aviation squadrons and, 35–36; aviation technical school and flight-training of, 10; Bulgarian navy and, 28; Central Powers' attack response by, 65–66; demobilization (1918), 123; flight training for Army versus, 18–19; Grand Duke Aleksandr and, 10; Great War aircraft of, 2, 19; *morskoi* aircraft and, 12; Sevastopol' Aviation School and, 20; Stavka and, 50
inflation, 108, 110
Institute of Aerodynamics, 16
Inter-Allied Aviation Inspection Committee, 188
International Women's Day protests (1917), 108
Irben Strait, Gulf of Riga and, 73
Irina, Princess, 92
Irmingard, 103
Istrebitel' No. 1 (I-1) and No. 2 (I-2), 181–82, 184, 197
Italy, 1, 26–27, 91, 139
Iudenich, Nikolai N., 158, 159–60
Iusupov, Feliks, Prince, 92

Japan, 2–3, 128, 172–73
Jesionowski, Kazimierz, 167
Ju-20 floatplane, 189–90
Ju-21 floatplane, 190
Jullien, Pierre, 7
Junkers, Hugo, 187
Junkers Flugzeuwerke AG, 187–88

Kagul, 29, 34
Kaiserin (Germany), 122
Kalinin, Konstantin A., 133
Kanin, Vasilii A., 74–75
Kasatkin, Vladimir, 94–95
Kashalot (Russian submarine), 106
Katia, 78
Kerenskii, Aleksandr F., 114, 116–17, 126–27
Khanzhin, Mikhail V., 149–50
Khodorovich, Viktor A., 22–23
Kiev: pilot training at, 23; pre–Great War aviation park at, 35
King David, 48
Kit (Russian submarine), 106
Klembovskii, Vladislav N. K., 116
Kluck, Alexander von, 37
Knox, Alfred W. F., 82, 146
Kolchak, Aleksandr V.: Belousovich and Modrakh transfer to, 147; as Black Sea Fleet commander, 105, 112; capture and execution of, 152–53; defeat of, 148; Red Air Force and Army north of Ufa and, 151–52; as White ground forces head, 145–46
Komendantskii Aerodrome, 53–54
Konarmiia (First Red Cavalry Army), 167–68
König Albert (Germany), 122
Konstenchik, Avenir M., 93–95
Kornilov, Lavr G., 129, 130, 158
Kornilov Division, 158, 160
Korpusnoi Aerodrome, Saint Petersburg, 28–29
Kostkin, Ivan M., 182
Kovan'ko, Aleksandr A., 156, 163, 171, 175
Kozakov, Aleksandr A.: as ace pilot, 61, 70–71, 101–2, 114–15; British military mission welcomes, 139; death of, 147; scouting over Sixth Red Army, 143; SPAD S.7 fighters for, 107; varied surname spellings, 207n27
Kozhevnikov, Ivan I., 156
Krasin, Leonid B., 161, 162, 189
Kronshtadt Fortress uprising (1921), 175–76, 196
Kruten', Evgraf N., 70, 114–15
Kuban Cossack region, 129, 170–71
Kubanets, 104
Kul'tin, Leonid A., 150–51, 167
Kuropatkin, Aleksei N., 100

Langley, Samuel Pierpont, 3
Latvia, Soviet peace treaty with, 160
Lavrov, Georgii I., 13, 54–55, 99, 113
Le Rhône engines, 15, 63–64, 141
Lebedev Aeronautics Company: aircraft production (1916) by, 90; FBA flying boats built under license by, 75; formation of, 11; GAZ-1 and end of operations at, 180; Lebed XII (Swan) airplane of, 13; Soviet-controlled Dukh factory and, 144–45
Left Socialist Revolutionaries, 127–28
Lenin (Vladimir I. Ulyanov/Ulianov): all-Bolshevik cabinet of, 127; Allied troops and planes in North Russia and, 141; on All-Russian Board for the Administration of the Air Fleet, 130; armed forces goals of Trotskii versus, 152; on Kolchak's execution, 153; New Economic Policy of, 176; on peace treaty with Germany and Central Powers, 125; Revolutionary Military Council chaired by, 214n20; Russian Civil War and, 126; shocked by White victories, 150; TsAGI and, 133, 178; on White movement destruction, 129; wounding of, 128; on Zhukovskii, 16
Leopold I, Bulgarian Tsar, 27
Levavasseur, Léon, 9
Libau, German attack on, 64–65, 72, 207n7
Liberty engine, 14, 181, 182
Libya (formerly Tripolitania), 26–27
Lida (now in Belarus), pre–Great War aviation park at, 35

INDEX

Lindbergh, Charles, 9
Lithuania, Soviet peace treaty with, 160
Litvinov, Maxim, 207n5
Litvinov, Vladimir A., 75, 76, 207n5
Lobov, Vladimir, 98–99
Loiko, Ivan A., 115
Lowicz, Russians bombing German railroad station at, 70
Ludendorff, Erich F. W., 43–44, 71–72, 78
Luft-Verkehrs Gesellschaft (LVG), 62

M-9 (Grigorovich flying boat), 11–12, 74, 95, 96–98
Macedonia, 26, 27, 78–79
machine guns: for aircraft, 7, 61; Cossack-dominated Russian Army, 171; on D. XI, 185; on D.H. 9 with Scarff Ring, 140; empty Russian supply depots and, 68; for FBAs (1915), 74; fixed, with interrupter gear, 144; in front of pusher-type aircraft, 100, 102; German, against Seversky, 76; German, upward-angled, 64; German forward-firing, 111; Hotchkiss, 62, 63; on I-2, 181; on Il'ia Muromets, 61, 80–81, 83, 92–93, 99; imported, 63, 153; low-flying Russian planes hit by, 44–45; Madsen, 63, 80, 81; Maxim, 61, 63, 74, 102, 182; mounted, 70; on R-1, 181; on R-02, 190; on R.E. 8, 140; reconnaissance planes (1914) and, 50; Red Air Fleet, 174; Red Air Force, 151–52, 160, 168, 170, 173; rotating, 119; of Russian fighters (1917), 63–64; Russian shortage of, 90; on S-20, 181; Seversky's air battle with Germans and, 96–98; Soviet military, 129–30; swivel-mounted, 63; synchronized with propellers, 13, 62, 63, 74, 194; Vickers, 107, 140, 144, 155, 181; White Army, 164, 165
Mackensen, August von, 44, 67, 78, 105
Magdeburg (Germany), 37–38
Mai-Maevski, Vladimir Z., 154, 157
Makhsheiev, Dimitrii K., 99
Makienok, Donat A., 114–15
Malina, Franz, 40
Mamontov, Konstantin K., 157, 158
Mannerheim, Carl Gustaf Emil von, 124

Marne, Battle of the, 38
Marx, Karl, 127
Marxists, Soviet of Workers' Deputies and, 109
Masaryk, Tomáš G., 138
Matson, Garaľd A., 173–74
Mawdsley, Evan, 126
mechanics. *See* aircraft repairs and support personnel
Meller, Iurii A., 11. *See also* Dukh Company
Mel'nikov, Nikolai S., 174
Mesheraup, Petr, 170
Mikulin, Aleksandr A., 133
Military Aviation School, 156
military hardware manufacturing, Alekseev and, 90
Military Revolutionary Committee (MRC), 126–27
Miller, Evgenii K., 147
Millerand, Alexandre, 169
Mirbach, Wilhelm von, 128
Modrakh, Sergei K., 139, 147–48, 164
Mokotov Aerodrome, 166–67
Moltke, Helmuth von, 38, 43
Mongolfier, Joseph and Jacques, 4–5
Montenegro, Balkan League and, 26–27
Moon Island (Baltic), 64, 74, 119
Morane-Saulniers aircraft, 11, 12, 29, 63, 71
morskoi (naval) aircraft, 11–12
Morzh (Russian submarine), 106
Moscow: aircraft industry in, 15, 189, 198; flight training at, 23, 101, 111, 112, 118, 131; pre–Great War aviation park at, 35. *See also* Soviet Russia
Moscow School of Theoretical Aviation, 17, 178
motors. *See* engines, aircraft
Motovilikha Artillery Works, Perm, 149
Mozhaiskii, Aleksandr F., 3–4
Murav'ev, Mikhail A., 137
Murmansk, 125, 139, 140–41, 147

Nadezhda (Bulgarian gunboat), 79
Nadezhdin, Vadim M., 154, 163
Narval (submarine), 106
Naumov, Aleksandr A., 164, 175
Nepenin, Adrian I., 111, 112, 211n13

Nerpa (submarine), 106
Nesterov, Petr N., 24, 39–40, 41, 71
New Economic Policy (1921), 176, 180
Nieman Army, 71–72, 84
Nieuport, Alfred and Charles de, 24
Nieuports: air battle with D.H. 9 (1920), 170; Anatra models, 12; Dybovskii's flying of, 24; First Slavic-British Aviation Squadron and, 140; importing difficulties, 48; land-based pursuit training in, 111; with machine gun synchronization, 63; obsolescence issues, 47; Polikarpov and, 180; Red Air Force flying, 155; for Russian combat, 193; as Russian license-built aircraft, 11, 24–25, 29; Sergievsky shooting down balloons in, 7
Nikolai (hydrocruiser), 65, 77, 79, 103, 106
Nikolai II, Tsar: abdication of, 108; air fighter detachments created by, 100; Aleksandr on rumors about, 92; on balloons in Vladivostok, 5–6; on Dmitriev as Twelfth Army commander, 82; flight in *Grand* with Sikorsky and, 54; military command by, 84; pilot awards by, 24, 40, 51, 55, 56, 98; Second Balkan War and, 27; Sevastopol' Aviation School and, 21; Shidlovskii's commissioning and, 59; Western technology and, 2
Nikolai Nikolaevich, Grand Duke, 35, 84
Nilufer (Turkey), 50
Nizhevskii, Robert L., 85–86
North Russia: air superiority over Soviets in, 143; Allies' challenges in, 146–47; British and U.S. soldiers in, 128; Moscow's presence in, 139; during Russian Civil War, *134*; Slavic-British Aviation Corps in, 132; Soviet propaganda distributed in, 177
Northwestern Army, White, 158–60
Northwestern Front, Russia's: diversionary attack against Germans on, 116; offensives faltering on (1916), 90; Orlov and defense of, 88; reconnaissance at, 58; reduction of German soldiers on, 78; revolution and continued offensives along, 114; Russian offensive failure at, 100

Novik, 37
Novogeorgievsk Fortress, 6, 87

Occidental Petroleum Corporation, 187
Odensholm (Baltic), 37
Odessa: pilot training at, 23; pre-Great War aviation park at, 35; radical soviets in (1917), 112; Turks attack boats in harbor of, 50
Odessa Aero Club, 12–13, 41
Omsk, Siberia: Directorate in, White military forces and, 148–49; pilot training at, 23; to Ufa, during Russian Civil War, *135*
Order No. 1, Soviet, 109, 194
Orlitsa (hydrocruiser), 64, 73
Orlov, Ivan A., 13, 86, 87–88, 114–15
Ösel Island (Baltic): FBAs flying out of, 73–74; German mines laid around, 119; Nieuports for squadrons on (1917), 119; *Orlitsa* operations near, 64, 74; Russian naval air stations on, 37; Second Bombing-Reconnaissance Squadron at, 75–76
Osoaviakhim (Association for the Promotion of Defense, Aviation, and Chemical Warfare), 191
Ottoman Empire: amphibious operations against, 104–5; German friendship with, 49; Russia declares war (1914) on, 50–51; Russian conflict with (1877–1878), 5; Russian task force in Black Sea and, 65; Serbian control by, 1. *See also* Turkey
Ozerskii, Dimitrii A., 82–83, 84–85

Pallada (cruiser), 37, 65
Pamiat' Merkuriia (cruiser), 106
Panasiuk, Vladimir D., 54–55
Pankrat'ev, Aleksei V., 58, 85–86, 100, 118, 155
Petkevick, Evgenii I., 174
Petr the Great, Tsar, 2
Petrograd (Saint Petersburg): flight training at, 17, 131; Iudenich and Northwestern Army toward, 158–60; land and freedom speech by (March 1917), 113; Leningrad as new name

for, 131; pilot training at, 23; pre–
  Great War aviation park at, 35;
  Russian government in, 1; Special
  Russian Aviation Squadron from, 27
Petrograd Telegraph Agency, 109
pigeons, Russian balloon–ground
  communication using, 6
pilots: candidates, class status and, 21, 23,
  24, 26, 29, 101, 131–32; oath of service
  for, 109, 194–95; tsarist, Aviation
  Board on, 131–32. *See also* flight
  training
Pilsudski, Jozef, 166, 167, 168
Pishvanov, Aleksandr M., 115, 163
Platz, Reinhold, 155, 219n16
Plevna siege (1877–1878), 5, 6
Pliat, Marcel, 94–95
Poland: aircraft overhauls at Mokotov
  Aerodrome, 167; ambitions for
  eastern boundary of, 153; Brest-
  Litovsk Treaty and, 126; Soviet
  Russian territories invaded by,
  162–63, 166, 167–69; Vrangel' seeking
  Russian Army support from, 172–73;
  Whites' offensive from Crimea against
  Red Army and, 165
Polikarpov, Nikolai Nikolaevich, 180–81,
  182–83, 184, 197
Politbureau, aviation industry and, 179
Polupanov, Aleksandr V., 137
Popov, Aleksandr A., 182
Portsmouth, Treaty of (1905), 3
Princep, Gavrilo, 1
Prittwitz und Gaffron, Maximilian von, 43
Products Exchange Company (Prodexo),
  179–80
Pronzitel'nii (Russian destroyer), 77
propellers, wooden, Petrograd companies
  manufacturing, 16
Provisional Government: Czech Legion
  recruiting in POW camps and, 138;
  demonstrations opposing, 114; end of,
  127; formation of, 108; Lenin on, 126;
  offensive against German and
  Austrian troops by, 195; respect lost
  for, 117; soviet support for, 113; U.S.
  line of credit to, 186. *See also*
  Kerenskii, Aleksandr F.
Prussis, Khristianskii F., 54–55

Przemysl Fortress, 66
Przhegodsky, Valerian, 110

R-1 (*razvedchik*—reconnaissance type-1)
  aircraft, 180
R-02 (*razvedchik*-No. 02) airplanes, 190
Ragozin, Nikolai, 29, 34
Rapallo, Treaty of, 187, 189, 198
Rasputin, Grigorii E., 92
R.E. 8s (aircraft), 140, 155
Rebikov, Nikolai V., 11
reconnaissance, air: for Balkan War,
  Special Russian Aviation Squadron
  and, 27; early aircraft and, 10; Essen
  on airplanes in Baltic Sea for, 64; by
  EVK off Gulf of Riga coast, 93, 95;
  Gatchina training for, 23; by Germans,
  over Warsaw, 62–63; by Germans
  against Russians, 61; Great Retreat
  support by, 69–70; Ju-20 floatplane
  for, 189–90; manned balloons for, 4–5;
  Orlov and, 87–88; by Poland against
  Red Army, 166; Red Air Force, 151,
  152; Russian, in Baltic Sea (1914), 37;
  Russian limitations at start of Great
  War, 44–45; Russian shortage of
  artillery shells and, 67; Russians on
  balloons versus dirigibles for, 7; from
  Stan'kovo Aerodrome by Il'ia
  Muromets aircraft, 99; Sukhomlinov
  on airplanes and, 25; by Tkachev
  against the Austrians, 42; Tomson
  and, 88; of Turkish torpedo boats near
  Odessa, 50; Twenty-First Corps Air
  Squadron and, 70
Reconnaissance Experimental #8 (R.E. 8)
  aircraft, 140
Red Air Fleet, 173, 174
Red Air Force: aircraft of (1919), 145;
  counteroffensive against Poles and,
  168, 169–70; counteroffensive against
  Whites and, 150; evicts Whites from
  Orel, 160; Kronshtadt Fortress
  uprising and, 176; *Osoaviakhim*
  preparing youth for service in, 191;
  Petrograd defense from Whites and,
  159–60; proposals for indigenous
  fighter aircraft for, 181–82; Russian

Army Air Force strength versus, 173–74; Russian Civil War battle operations of, 177; Sopwith 1 ½ Strutters of, 151; special aviation group of, 157–58; against Whites north of Ufa, 151–52

Red Army: disagreements on direction for, 152; evicts Whites from Orel, 160; First, counteroffensive against Whites and, 150; formation of, 123; Kronshtadt Fortress uprising and, 176; Polish invaders and, 167–68; Seventh, Petrograd defense from Whites and, 159–60; Sixth, Allied troops and planes in North Russia and, 141–42; taking and occupying Lithuanian Peninsula, 174–75; Tenth, Civil War and, 156; Third, occupying territory won from Whites, 152; Volunteer Army against, 130; White forces and, 145; Whites' offensive from Crimea against, 165–66; Worker and Peasant, 133, 141

Red Fleet, formation of, 123

Red Guards, 129–30, 137

Reds: advantage over Whites, 145. *See also* Russian Civil War; Soviet Russia

Renault engines, 15, 16

Rennenkampf, Pavel, 43–44

Reparations Commission, 186

replacement parts. *See* aircraft repairs and support personnel

retreat. *See* Great Retreat

Riaboff, Aleksandr, 110

Riga: German submarine port at, 120; Russian defense against Germans taking, 37, 72

Riga, Gulf of: EVK reconnaissance off coast of, 93, 95; German armada takes, 120–22, 195; Russian warships assigned to, 207n7; Seversky's squadron protecting, 119–20; spring 1917 and naval officer/sailor relationship in, 111. *See also* Angern, Lake

Riga, Treaty of, 169

Romania, joins Allies in the War, 105

Romanian front, 114, 115

Romanov, Vladimir A., 158

Romanov family. *See* Aleksandr Mikhailovich, Grand Duke; Aleksandra Feodorovich; Nikolai II, Tsar; Nikolai Nikolaevich, Grand Duke

Romanovskii, Ivan Pavlovich, 82, 83

Roosevelt, Theodore, 3

Rosenthal, Friedrich, 40

Rostislav (dreadnought), 104

Royal Air Force (RAF), 91, 140, 155, 161, 162

Royal Aircraft Factory (British), 140

Rudnev, Evgenii V., 21–22, 56–57, 154, 175

Runo Island (Baltic), 95

Russia: aircraft engine difficulties for, 14–16; aircraft purchases for, 10, 24; aviation development claims by, 14; Baltic Sea region during the Great War, *33*; combat-ready aircraft (spring 1917), 107; debts from Great War, 186; dirigibles and, 7–8; eastern borders of Germany and Austria with, 30; foreign imports for war by, 2, 47–48, 90, 91; gliders and, 9; Great Retreat, *32*, 69–70, 71, 73–88; Great War aircraft manufacturers, 7, 10–14; Great War and, 1; military balloons for, 5, 6–7; Mozhaiskii's 'flight' (1884), 3–4; peasant soldiers uneducated about air squadrons, 46–47; rampant inflation (spring 1917), 108; Russo-Japanese War (1904–1905) and, 2–3; wars against Ottomans and, 27; wartime planning gaps, 67; western fortresses of, 6. *See also* Imperial Russian Army; Imperial Russian Navy; Soviet Russia

Russia (Model XI monoplane aircraft copy), 11

Russian Air Force Museum, 211–12n22

Russian Army: Il'ia Muromets and, 117–18; quits fighting after armistice, 130; Ulagai's failed expedition into Kuban region and, 171. *See also* Vrangel', Petr Nikolaevich; White movement

Russian Army Air Force, 172, 173, 174

Russian Army in the World War, The (Golovine), 102

Russian Civil War (1918–1921): aviation and, 143–60; Crimea during, *136*; D. VII

INDEX

aircraft bought for, 185; Efimov and, 24; events starting, 126–27; North Russia during, *134*; from Omsk, Siberia to Ufa during, *135*; Reds (Communists) versus Whites (anti-Communists), 126; White movement aviation and, 196

Russian Eighth Army, 46, 66–67, 101–2, 206n18

Russian Eleventh Army, 66, 116

Russian Empire, 206n24

Russian Fifth Army, 93, 95

Russian First Army, 43–44, 71, 88

Russian Fourth Army, 99, 115, 116

Russian Orthodox priests, marriage by, 209n8

Russian Revolution (1905), 52

Russian Revolution (1917): decline in military discipline and, 194; Dukh Company fighter plane production and, 63–64; economic causes of, 90, 108, 110; Nieuports purchased for, 25; Reds versus Whites in, 123–42; Soviet victories in 1920 and 1921, 161–76; squadrons for Great War and, 106–22; Sukhomlinov's escape from Russia and, 67

Russian Second Army, 43, 44, 66, 69, 70, 88, 99

Russian Seventh Army, 100, 101

Russian Sixth Army, 115–16

Russian Tenth Army, 99, 115

Russian Third Army, 67–69, 71, 79, 82

Russian Twelfth Army, 9, 82, 84, 95, 116, 120

Russo-Baltic [Railroad] Wagon Company: aircraft engine production by, 15–16; aircraft production by, 90; factory turned over to Junkers, 188, 189–90; GAZ-1 and end of operations at, 181; Sikorsky and, 12, 13, 51. *See also* Sikorsky, Igor I.

Russo-Japanese War (1904–1905), 2–3, 4–5, 10

Rutkovskii, Viacheslav S., 150

Safonov, Ludmila, 124, 125, 126

Safonov, Mikhail I., 119, 121–26

Saint Evstafi, 51

Salmson engines, 15, 96

Samsonov, Aleksandr, 43–44, 66

Samsun, 65

Santos-Dumont, Alberto, 9

Sapozhnikov, Georgii S., 168

Sarajevo, Great War and assassination in, 1

Sarnavskii, Vladimir G., 20

Scheidemann, Sergei, 69, 70, 206n24

Schlieffen, Alfred von, 36

Schlieffen Plan, 36–37

Schmidt, Ehrhardt, 120–21

Schneider, Franz, 62

scouting missions. *See* reconnaissance, air

Second Bombing-Reconnaissance Squadron, 75, 95, 96

Semenov, Grigorii M., 149

Serbia: Austrian declaration of war on, 1, 55; Balkan League and, 26–27; Bulgaria and Germany to attack, 78–79; First and Second Balkan Wars and, 28; Vrangel' seeking Russian Army support from, 172–73

Serbrennikov, Anatolii A., 55

Sergeev, Aleksei V., 141

Sergievsky, Boris, 7, 206n18

Sevastopol': Grand Duke Aleksandr and military aerodrome at, 10; Turks attack boats in harbor of, 50

Sevastopol' Aviation School: broader vision of pilot training and, 25; Gatchina pilot candidates and, 26; opening of, 20–21; pilot training at, 22–23; Rudnev as pilot-officer at, 21–22; Tkachev training at, 41; Ukrainian government control of, 131

Seversky, Alexander: air battle with Germans, 96–98; bombing mission against Germans, 76–77, 96–97; Central Powers' peace treaty with Soviet Russia and, 213n5; early Russian military aviation and, 11; German armada to Ösel Island and, 121–22; Germans bombing Zerel air station and, 112; as Moscow Aviation School flight instructor, 118; moves to San Francisco, 124, 195; ordered to Moscow Aviation School, 111; pep talk to Shchetinin workers by, 118–19, 212n26; pilot training for, 22–23, 26; post–Great War survival, 123; Russian versus American

name for, 75; shoots down German sea plane in the Baltic, 95; Trotskii signs pass for travel to U.S. by, 125; at Zerel naval air station, 119–20
Seversky, George, 11
Seversky, Nicholas G., 11, 80
Shavrov, Vadim Borisovich, 14
Shchetinin, Sergei S., 27
Shchetinin Works, 11, 13, 77, 90, 118–19
Shebalin, Sergei K., 139
Shidlovskii, Mikhail V., 52–53, 59, 85, 113
Shirinkin, Andrei D., 168
Shishkevich, Mikhail V., 26
Shishmarev, Mikhail M., 181
Shmeur, Georgii N., 94–95
Shvetsov, Arkadi D., 190
Siberia: Frunze pursuing Whites into, 152–53; Russian pilots in North Russia and forces from, 140; White forces defeated in, 196; White movement emergence in, 138–39. *See also* Omsk, Siberia
Siberian Army, 150
Sidorin, Vladimir I., 154, 157
Sigrist, Fred, 155
Sikorsky, Igor I.: Baltic Sea action and planes by, 51–52; bomb's near miss of, 205n7; departs EVK in protest, 113; *Grand* construction and testing, 53–54; Il'ia Muromets redesign by, 55–56, 84, 94; Il'ia Muromets test flight and, 54–55, 205n27; migration to United States, 65, 195; pilot training for Il'ia Muromets and, 56–57; as Russian aviation pioneer, 12, 14, 193; Russo-Baltic Wagon Company aircraft and, 13; seaplanes, in the Baltic, 34; Shidlovskii and, 59; Utgof and, 34, 65
Sikorsky, Olga Fyodorovna Simkovich, 60
Sikorsky, Sergei, 65
Sikorsky, Tania, 60, 65
Sikorsky S-6, 29
Sikorsky S-10, 29, 51
Sikorsky S-11, 29
Sikorsky S-16, 13, 51–52
Sikorsky S-20, 180–81
Skliansk, Eduard M., 141
Skoropadsky, Pavel, 131, 137
Slashchev, Iakov, 162

Slava (cruiser), 122
Slavic-British Aviation Corps, 132, 140, 147; Squadrons, 140, 142, 143
Slowik, Karol, 167
Smirnov, Ivan V., 115
Smirnov, Mikhail V., 81
Smirnov, Sergei F., 141
Smith, Herbert, 155
Socialist Revolutionaries, 128, 137
Socialist Revolutionary Party, 127
Society of Friends of the Air Force (*Obshchestvo Druzei Vozdushnovo Flota*, ODVF), 190–91, 198
Somme, Battle of the, 89
Sopwith, Thomas O. M., 155
Sopwith 1 ½ Strutters, 144, 145, 151, 180, 181
Sopwith Camels, 154–55, 157
Sopwith Snipes, 143–44
Souchon, Wilhelm, 49, 51
Southwestern Front, Russia's, 101–3, 114–15, 116, 117. *See also* Kozakov, Aleksandr A.
Soviet currency, U.S. dollar and, 219n6
Soviet of Labor and Defense, 179
Soviet of Workers' Deputies, 108–9
Soviet Russia: aircraft development (1918–1924), 177–91; foreign concessionaires in, 187–88; Germany's collaboration on military aviation with, 185; new government of, on cease-fire with Central Powers, 117; Russia's debts from Great War and, 186; Treaty of Rapallo with Germany and, 187, 197–98. *See also* Bolshevik Party; Lenin; Red Army; Stalin, Iosif; Trotskii, Lev D.
Soviet Supreme Council of National Economy, 178
SPAD (Societe Pour L'Aviation et ses Derives) aircraft, 63, 102, 107, 155
Spasov, Mikhail P., 82, 85
Special Committee for Heavy Aviation, 178
Special Russian Aviation Squadron, 27
Squadron of Flying Ships. *See* EVK
Stalin, Iosif (Joseph), 131, 179
State Aviation Factory No. 1 (GAZ-1), 180
Stavka (Russian military headquarters): on amphibious operations, 106; on Black

sea strategy, 49–50; German offensive and relocation of, 84; Gorshkov's bombing mission and, 61; on Il'ia Muromets, 58; machine guns for EVK and, 63; on pilot promotions, 92; on Rennenkampf in East Prussia, 43; spring 1917 plans of, 107; supplying FBA flying boats to Baltic Fleet, 72
Steklov, Lieutenant, 96
Stepanov, Ivan P., 156
Stroev, Mikhail P., 157–58
Strzhizhevskii, Vladimir I., 115, 163
Suk, Grigorii E., 115
Sukhomlinov, Vladimir A., 25–26, 28, 29, 34, 47, 67
Sveshnikov, Aleksandr N., 139, 142
Svetukhin, Junior Lieutenant, 50

Tannenberg, Battle of, 44, 66
taran (Russian battering-ram maneuver), 39–40, 71
Tashkent, pilot training at, 23
Tekhtel', Karl, 160, 217n34
Thomas, George, 140
Timofeev, Anatolii K., 163
Tkachev, Viacheslav M.: air battle with Red Air Force group head, 170; career achievements, 41–42; Crimea evacuation and, 175; First Kuban Aviation Squadron and, 156; opposition to Soviet government by, 129; Vrangel' seeking foreign support for Russian Army and, 172–74; White Army aviation in the Crimea and, 162, 164
Tomson, Petr-Eduard M., 86–87, 88
Trans-Siberian Railroad, 48, 125, 138, 148, 149, 195
Trebizond, Turkey, 51
Trenchard, Hugh, 91
Trotskii, Lev D. (Trotsky, Leon): counteroffensive against Whites and, 150; on disarming or shooting Czech soldiers, 138; Gatchina aviation school name and, 131; Kronshtadt Fortress uprising and, 176; Petrograd's defenses and, 159–60; Second Congress of Soviets and, 127; Seversky and, 124–25;

Society of Friends of the Air Force formed by, 190–91; Soviets name Gatchina after, 160; as War Commisar under Lenin, 152, 153
TsAGI. *See* Central Aero-Hydrodynamic Institute
TsAGI Works, 178
Tu-2 twin-engine tactical bomber, 197
Tuchkov, Aleksandr A., 118, 124
Tukhachevskii, Mikhail N., 152, 168, 176
Tumanevskii, Aleksandr K., 174
Tunoshenskii, Ivan N., 156, 164
Tupolev, Andrei Nikolaevich, 178–79, 197
Turkestan Army, 150
Turkey, 1, 153. *See also* Ottoman Empire

U-1 training plane, 183
U-boats, 48, 77, 78, 120
Ufa: from Omsk, Siberia to, during Russian Civil War, *135*; White Western Army captures, 150
Ufimsk Operation, 150
Ukraine, 126, 137, 155
Ulagai, Sergei G., 170–71
Ulianov (Ulyanov), Vladimir Ilich, 16, 128. *See also* Lenin
Union of Soviet Socialist Republics (USSR), establishment of, 182
United States: aircraft engines exported to Russia (1916) by, 91; anger over Decree on Peace by, 128; Arcos branch opened in, 179–80; Kolchak as White movement leader and, 146; loans for Russia and, 68; Russian aircraft manufacturing and, 13–14; Russian embassy moves to Vologda, 139; Russian pilots in North Russia and, 139–40; Seversky's migration to, 124, 125–26; Vrangel' seeking Russian Army support from, 172–73; Whites' evacuation from Crimea and, 175
Urbabk Aerodrome, 157
Utgof, Viktor, 29, 34, 65–66

Vakulovskii, Konstantin K., 86, 87, 116
Variol, 48
Vasil'chenko, Nikolai N., 173

Versailles, Treaty of, 186, 188, 198
Vinnitsa, 107, 113, 117
Vladivostok, 5–6, 22, 48, 139
Voevodskii, Nikolai S., 62–63
Voisin, Charles and Gabriel, 10
Voisins (aircraft), 11, 12, 70
Volkhov Field, 4, 5, 6, 22
Vologda, British and U.S. embassies relocate to, 139
Vologodtsev, Lev K., 167
Volunteer Army, 129–30, 137, 154, 216n23. *See also* Denikin, Anton I.
Vrangel', Petr Nikolaevich: Armed Forces of South Russia and, 154; in Crimea, offensive against Red Army and, 165; Crimea evacuation under, 175; defending Tsaritsyn, 158; as Deniken's replacement, 163–64; Modrakh joins White forces under, 148; Polish-Soviet war (1920) and, 169, 170; Red Army at Tsaritsyn and, 157; Red troops at Velikokniazheskaia and, 156; seeking foreign support for Russian Army, 171–73; Tkachev as aviator in the Crimea under, 162; White partisan group under Kuznetsov and, 129

war, demonstrations against (1917), 114
War Ministry, 12, 25–26. *See also* Sukhomlinov, Vladimir A.
Warsaw, 6, 23, 35, 166–67
Western Front, Russia's, 115
White Guard (Finland), 124
White movement, 151–52; after Brest-Litovsk Treaty, 195; Allied Powers' hopes for German defeat with, 139; Army, reorganized, 164–65; control of major cities by, 158; Czech Legion and, 148–49; defeat at Petrograd, 159–60; French support during Polish-Soviet war (1920) for, 169; military arms and equipment for (1919), 153–54; in North Russia, 141–42; in North Russia, challenges for Allies and, 146–47; Red Army's overextension and, 155–56; Red Guards and, 129–30; Reds advantage over, 145; reorganization of Army and force (1920), 163–64; Russian pilots in North Russia and, 140; Western Army, 149–50. *See also* Armed Forces of South Russia

White Russia (now Belarus), 126
Wilhelm II, Kaiser, 49, 55
Wilson, Woodrow, 125, 139–40, 146
Worker and Peasant Red Military Air Fleet, 118, 133, 141
Worker-Peasant Red Military Air Force, 145. *See also* Red Air Force
World War I. *See* Great War
World War II, 196–97
Worm Island (Baltic), 64, 74
Wright, Orville and Wilbur (Wright Company), 3, 4, 9, 14, 18, 201n32
Wright Flyer, 3, 4
Wright Model C aircraft, 4

Zakharov, Anatolii T., 163, 175
Zegevol'd, EVK headquarters at, 93
Zenzinov, Vladimir M., 148
Zeppelin, Graf Ferdinand von, 8
Zerel, 73, 119, 120. *See also* Ösel Island
Zhkov, Andrei I., 182
Zhloba, Dimitrii P., 165
Zhukovskii, Nikolai E., 16, 17, 75, 178–79, 197
Zhukovskii Military Air Academy, 178, 197
Zinoviev, Grigorii E., 159
Zvegintsev, Lieutenant Colonel, 83
Zverev, Fedor T., 156, 163, 175

Sea strategy, 49–50; German offensive and relocation of, 84; Gorshkov's bombing mission and, 61; on Il'ia Muromets, 58; machine guns for EVK and, 63; on pilot promotions, 92; on Rennenkampf in East Prussia, 43; spring 1917 plans of, 107; supplying FBA flying boats to Baltic Fleet, 72
Steklov, Lieutenant, 96
Stepanov, Ivan P., 156
Stroev, Mikhail P., 157–58
Strzhizhevskii, Vladimir I., 115, 163
Suk, Grigorii E., 115
Sukhomlinov, Vladimir A., 25–26, 28, 29, 34, 47, 67
Sveshnikov, Aleksandr N., 139, 142
Svetukhin, Junior Lieutenant, 50

Tannenberg, Battle of, 44, 66
taran (Russian battering-ram maneuver), 39–40, 71
Tashkent, pilot training at, 23
Tekhtel', Karl, 160, 217n34
Thomas, George, 140
Timofeev, Anatolii K., 163
Tkachev, Viacheslav M.: air battle with Red Air Force group head, 170; career achievements, 41–42; Crimea evacuation and, 175; First Kuban Aviation Squadron and, 156; opposition to Soviet government by, 129; Vrangel' seeking foreign support for Russian Army and, 172–74; White Army aviation in the Crimea and, 162, 164
Tomson, Petr-Eduard M., 86–87, 88
Trans-Siberian Railroad, 48, 125, 138, 148, 149, 195
Trebizond, Turkey, 51
Trenchard, Hugh, 91
Trotskii, Lev D. (Trotsky, Leon): counter-offensive against Whites and, 150; on disarming or shooting Czech soldiers, 138; Gatchina aviation school name and, 131; Kronshtadt Fortress uprising and, 176; Petrograd's defenses and, 159–60; Second Congress of Soviets and, 127; Seversky and, 124–25;

Society of Friends of the Air Force formed by, 190–91; Soviets name Gatchina after, 160; as War Commisar under Lenin, 152, 153
TsAGI. *See* Central Aero-Hydrodynamic Institute
TsAGI Works, 178
Tu-2 twin-engine tactical bomber, 197
Tuchkov, Aleksandr A., 118, 124
Tukhachevskii, Mikhail N., 152, 168, 176
Tumanevskii, Aleksandr K., 174
Tunoshenskii, Ivan N., 156, 164
Tupolev, Andrei Nikolaevich, 178–79, 197
Turkestan Army, 150
Turkey, 1, 153. *See also* Ottoman Empire

U-1 training plane, 183
U-boats, 48, 77, 78, 120
Ufa: from Omsk, Siberia to, during Russian Civil War, *135*; White Western Army captures, 150
Ufimsk Operation, 150
Ukraine, 126, 137, 155
Ulagai, Sergei G., 170–71
Ulianov (Ulyanov), Vladimir Ilich, 16, 128. *See also* Lenin
Union of Soviet Socialist Republics (USSR), establishment of, 182
United States: aircraft engines exported to Russia (1916) by, 91; anger over Decree on Peace by, 128; Arcos branch opened in, 179–80; Kolchak as White movement leader and, 146; loans for Russia and, 68; Russian aircraft manufacturing and, 13–14; Russian embassy moves to Vologda, 139; Russian pilots in North Russia and, 139–40; Seversky's migration to, 124, 125–26; Vrangel' seeking Russian Army support from, 172–73; Whites' evacuation from Crimea and, 175
Urbabk Aerodrome, 157
Utgof, Viktor, 29, 34, 65–66

Vakulovskii, Konstantin K., 86, 87, 116
Variol, 48
Vasil'chenko, Nikolai N., 173

Versailles, Treaty of, 186, 188, 198
Vinnitsa, 107, 113, 117
Vladivostok, 5–6, 22, 48, 139
Voevodskii, Nikolai S., 62–63
Voisin, Charles and Gabriel, 10
Voisins (aircraft), 11, 12, 70
Volkhov Field, 4, 5, 6, 22
Vologda, British and U.S. embassies relocate to, 139
Vologodtsev, Lev K., 167
Volunteer Army, 129–30, 137, 154, 216n23. *See also* Denikin, Anton I.
Vrangel', Petr Nikolaevich: Armed Forces of South Russia and, 154; in Crimea, offensive against Red Army and, 165; Crimea evacuation under, 175; defending Tsaritsyn, 158; as Deniken's replacement, 163–64; Modrakh joins White forces under, 148; Polish-Soviet war (1920) and, 169, 170; Red Army at Tsaritsyn and, 157; Red troops at Velikokniazheskaia and, 156; seeking foreign support for Russian Army, 171–73; Tkachev as aviator in the Crimea under, 162; White partisan group under Kuznetsov and, 129

war, demonstrations against (1917), 114
War Ministry, 12, 25–26. *See also* Sukhomlinov, Vladimir A.
Warsaw, 6, 23, 35, 166–67
Western Front, Russia's, 115
White Guard (Finland), 124
White movement, 151–52; after Brest-Litovsk Treaty, 195; Allied Powers' hopes for German defeat with, 139; Army, reorganized, 164–65; control of major cities by, 158; Czech Legion and, 148–49; defeat at Petrograd, 159–60; French support during Polish-Soviet war (1920) for, 169; military arms and equipment for (1919), 153–54; in North Russia, 141–42; in North Russia, challenges for Allies and, 146–47; Red Army's overextension and, 155–56; Red Guards and, 129–30; Reds advantage over, 145; reorganization of Army and force (1920), 163–64; Russian pilots in North Russia and, 140; Western Army, 149–50. *See also* Armed Forces of South Russia
White Russia (now Belarus), 126
Wilhelm II, Kaiser, 49, 55
Wilson, Woodrow, 125, 139–40, 146
Worker and Peasant Red Military Air Fleet, 118, 133, 141
Worker-Peasant Red Military Air Force, 145. *See also* Red Air Force
World War I. *See* Great War
World War II, 196–97
Worm Island (Baltic), 64, 74
Wright, Orville and Wilbur (Wright Company), 3, 4, 9, 14, 18, 201n32
Wright Flyer, 3, 4
Wright Model C aircraft, 4

Zakharov, Anatolii T., 163, 175
Zegevol'd, EVK headquarters at, 93
Zenzinov, Vladimir M., 148
Zeppelin, Graf Ferdinand von, 8
Zerel, 73, 119, 120. *See also* Ösel Island
Zhkov, Andrei I., 182
Zhloba, Dimitrii P., 165
Zhukovskii, Nikolai E., 16, 17, 75, 178–79, 197
Zhukovskii Military Air Academy, 178, 197
Zinoviev, Grigorii E., 159
Zvegintsev, Lieutenant Colonel, 83
Zverev, Fedor T., 156, 163, 175

ABOUT THE AUTHOR

James K. Libbey is professor emeritus at Embry-Riddle Aeronautical University, where he taught aviation history and Russian-American relations. His numerous publications include articles in such journals as *Aviation History, American Aviation Historical Society Journal,* and *Russian History,* as well as aviation-related entries in reference works such as *Air Warfare.* Six of Libbey's seven previous books deal with Russian personalities or the economic and political relations between the United States and Russia.

The Naval Institute Press is the book-publishing arm of the U.S. Naval Institute, a private, nonprofit, membership society for sea service professionals and others who share an interest in naval and maritime affairs. Established in 1873 at the U.S. Naval Academy in Annapolis, Maryland, where its offices remain today, the Naval Institute has members worldwide.

Members of the Naval Institute support the education programs of the society and receive the influential monthly magazine *Proceedings* or the colorful bimonthly magazine *Naval History* and discounts on fine nautical prints and on ship and aircraft photos. They also have access to the transcripts of the Institute's Oral History Program and get discounted admission to any of the Institute-sponsored seminars offered around the country.

The Naval Institute's book-publishing program, begun in 1898 with basic guides to naval practices, has broadened its scope to include books of more general interest. Now the Naval Institute Press publishes about seventy titles each year, ranging from how-to books on boating and navigation to battle histories, biographies, ship and aircraft guides, and novels. Institute members receive significant discounts on the Press' more than eight hundred books in print.

Full-time students are eligible for special half-price membership rates. Life memberships are also available.

For a free catalog describing Naval Institute Press books currently available, and for further information about joining the U.S. Naval Institute, please write to:

Member Services
U.S. Naval Institute
291 Wood Road
Annapolis, MD 21402-5034
Telephone: (800) 233-8764
Fax: (410) 571-1703
Web address: www.usni.org